The Politics of Women's Biology

The Politics

OF

Women's Biology

RUTH HUBBARD

Rutgers University Press
New Brunswick, New Jersey

Second paperback printing, 1992

Library of Congress Cataloging-in-Publication Data

Hubbard, Ruth, 1924-
The politics of women's biology / by Ruth Hubbard.
p.   cm.
Bibliography: p.
Includes index.
ISBN 0-8135-1489-4 (cloth)   ISBN 0-8135-1490-8 (pbk.)
1. Women—Physiology—Political aspects.   2. Human biology—Sex
differences—Political aspects.   3. Human reproductive technology—
Political aspects.   4. Feminist criticism.   I. Title.
QP81.5.H83   1990
305.42—dc20
                                                                89-10242
                                                                CIP

British Cataloging-in-Publication information available

Contents

Acknowledgments   ix
Introduction: Science and Science Criticism   1

Part One. How Do We Know?   7
  1. Science in Context   9
  2. Fact Making and Feminism   23
  3. Women in Academia   35
  4. Women Scientists   43
  5. The Double Helix: A Study of Science in Context   48

Part Two. What Do We Know?   67
  6. Genes as Causes   70
  7. Have Only Men Evolved?   87
  8. Human Nature   107
  9. Rethinking Women's Biology   119
  10. The Social Construction of Sexuality   131
  11. Constructing Sex Difference   136

Part Three. How Do We Use It?   141
  12. Medical, Legal, and Social Implications of Prenatal Technologies   147
  13. Prenatal Technologies and the Experience of Childbearing   161
  14. Who Should and Who Should Not Inhabit the World?   179
  15. Of Embryos and Women   199

Some Final Thoughts   209
Bibliography   213
Index   223

Acknowledgments

This book reflects the development of my ideas during almost two decades. Hence I cannot possibly acknowledge my debt to all my colleagues, students, friends, and family members who encouraged, pushed, criticized, and otherwise helped me. I thank all of you and hope that you will feel comfortable with the ways I have incorporated insights at which we may have arrived together.

I should, however, like to acknowledge two friends who have died all too soon and who therefore cannot receive their share of my collective thanks. One is Lila Leibowitz, anthropologist, feminist, friend, whose extensive knowledge, intelligent criticisms, and wonderful sense of humor made every joint endeavor a pleasure. The other is Larry Hill, chaplain in the United Ministry at Harvard and a comrade in many political struggles at Harvard, in Cambridge, and in the country. In 1974, Larry invited me to participate in a lecture series he was organizing for Harvard undergraduates and speak on the topic "Science as a Belief System." In the course of preparing this talk I began to admit to myself that I was becoming less interested in doing science than in thinking critically about its ideological basis, its social structure, and its political significance. I hope Lila and Larry would like this book and would be glad they had a part in its beginnings.

I would like to thank Nancy Newman for preparing the index and Trisha Dahill, Karen McCree-Diaz, and Elli Valminuto for help with typing the manuscript.

Earlier versions of some of the chapters in this book have appeared in *Biological Woman—The Convenient Myth,* R. Hubbard, M. S. Henifin, and B. Fried, eds. (Cambridge, Mass.: Schenkman, 1982); *Harvard Educational Review* ("Review of *Women Sci-*

entists in America: Struggles and Strategies to 1940 by Margaret Rossiter," 1984, 54 [4]:465–468); *Hypatia* ("Science, Facts, and Feminism," 1988, 3 [1]:5–17); *Ideology of / in the Natural Sciences,* H. Rose and S. Rose, eds. (Cambridge, Mass.: Schenkman, 1980); *International Journal of Health Services* ("Eugenics and Prenatal Testing," 1986, 16:227–242); *New Literary History* ("Constructing Sex Difference," 1987, 19:129–134); *Structures of Matter and Patterns in Science,* M. Senechal, ed. (Cambridge, Mass.: Schenkman, 1980); *Test-Tube Women,* R. Arditti, R. Duelli Klein, and S. Minden, eds. (London: Pandora Press, 1984); *Towards a Liberatory Biology,* The Dialectics of Biology Group, S. Rose, ed. (London: Allison and Busby, 1982); *Women Look at Biology Looking at Women,* R. Hubbard, M. S. Henifin, and B. Fried, eds. (Cambridge, Mass.: Schenkman, 1979); and *Women's Studies Quarterly* ("Reflections of a Feminist Biologist on Human Sexuality and Reproduction," 1984, 12 [4]:2–5).

The Politics of Women's Biology

Introduction:Science and Science Criticism

Nature is part of history and culture, not the other way around. Sociologists and historians of science tend to know that. Most scientists do not. Because I was trained as a scientist, it has taken me many years to understand that "in science, just as in art and in life, only that which is true to culture is true to nature" (Fleck, [1935] 1979, p. 35).

From 1947 until the late 1960s I was a devout scientist. I did experiments and wrote papers and reviews in the accepted tradition and did not ask myself how what I was doing fit into the culture. Like many scientists, I assumed I was probing nature and that that was an unquestionable good and reason enough to go on doing it.

The Vietnam War and the women's movement led me to look closely at these assumptions. The war made me see that science and the universities help maintain differences in wealth and power between nations and between the ethnic, racial, and economic groups within them. The women's liberation movement sharpened and focused these issues and gave them special urgency, for in the mid-1970s, no doubt in response to the renewed activism for women's rights, biological theories began to be revived that threaten women's struggle for equality. Sociobiologists and other biological determinists breathed new life into old arguments that derived differences in the positions women and men occupy in the workplace, the home, and the political sphere from the differences in our procreative functions. They needed to be answered.

That I was able to turn my attention to these issues was due to the fact that in 1973, owing in large part to the political work of the women's movement, the tenuous position I had held at Harvard became stable. In an unusual step, the university promoted a few of us from the typical women's ghetto of "research associate and lecturer" to tenured professorships. This promotion carried with it increased freedom to decide what work I wanted to do and what I wanted to teach. I could pursue my changing interests and share my new questions and analyses with a growing group of interested and interesting students. I could develop a network of colleagues from a secure professional and economic base.

This book describes my journey from observing nature to observing science, from doing science to studying it. The book is divided into three parts: "How do we know?" "What do we know?" "How do we use it?" The first is concerned primarily with feminist issues in the sociology of science, the second with feminist criticisms of subject matter, the third with current applications of biological knowledge in procreative technologies. As might be expected, this arrangement is to some extent arbitrary and involves cross-currents and overlaps.

I want to stress at the outset that I think the subject of women's biology is profoundly political—hence the book's title. The predominantly male scientists who have described our biology have done so at least in part to explain why it is "natural" for us to function as we do in society. To assert power over our lives women, of course, need the economic and political conditions that make it possible even to conceive of doing so. But we also need to recognize our biological capabilities in order to value them and use them to benefit us.

To go beyond defining ourselves as victims of male power and domination, we have to acquire the sense that our individual histories and needs, as well as our collective experiences and actions, are important. We need not only to accept but to appreciate our bodies and bodily functions, which is difficult to do in the face of the barrage of disabling information we encounter at home, in the media, and from health professionals.

Women's health activists have tried to reeducate us about our bodies. But to free ourselves from debilitating misrepresentations, we need to understand the ideological bases of the medical/scientific misinformation and disinformation we get about how genes, hormones, muscles—in a word, our bodies—function. To reconceptualize our biology and make it truly ours is important the same way it has been important for us to gain control of our language and interpret our own experiences. This book is intended to contribute to that process.

It seems contradictory that as our political awareness has increased, many of us scientists who are feminists have turned from doing scientific experiments to the social studies of science and science criticism. Feminist poets, novelists, and artists by and large have not needed to make such a change. They have been more able to incorporate their political consciousness into their work than we have.

I am not sure of all the reasons why this is so. One surely is that the way scientists define objectivity renders the various guises in which politics enter into scientific explanations invisible. For example, social class has never been a category in U.S. health statistics, and recently the Thatcher government has ceased to specify social class in British health statistics. The U.S. statistics are usually expressed in the quasi-biological terms of race and sex, which obscure the economic and social differences between the different races and sexes. We will encounter other examples as we go along. The important point for now is that political and social realities can be incorporated in subtle ways so that they are hard to discern in what are presented as descriptions of biology or, in this instance, health.

Another reason is that doing science requires institutional support. In literature and the arts, the most original work is not produced in universities or subject to review by funding agencies. It may require a sponsor or a publisher, but in the arts it is possible to find someone who will take the risk to publish or exhibit innovative work or, if need be, to form alternative collectives that take that risk. Probably, also, the rigid training scientists undergo limits our own imaginations and constrains our originality more than happens to artists.

It is appropriate to raise these issues because in science, as in the arts, we create stories and images about nature, including people, and transmit them through language. And, in both, the stories must be "true" in the sense that they must reflect other people's experiences. A writer who misrepresents people's life experiences is as "wrong" as a scientist who misrepresents or misinterprets cultural beliefs about nature.

At present, some social critics of science argue that, much like writers, all scientists can do is tell stories, and that, as in literature, these stories are grounded in the social and political realities of our time. This view is true in the sense that scientists do not simply go out and look at nature or hold up a mirror to it. They define and isolate the pieces of nature they choose to look at and, in so doing, change them by removing them from their natural context. They thus construct the nature they describe in their science. For laboratory scientists this process goes even further because they must start by constructing the "natural objects" they study.

There are many examples of the grounding of scientists' stories in the culture. Euro-American biologists have arranged the groups into which they have classified animals and plants into hierarchies, like the societies in which they have lived. They speak of the animal and plant "kingdoms," of "higher" and "lower" organisms. And, of course, they rank human beings at the top. Darwin's evolutionary theory is based on the idea that the history of organisms has unfolded in a world of scarce resources where individuals have to compete. In this world the individual is the most important unit, far more important than the group. Both assumptions reflect basic beliefs of Darwin's class and society. You will find other examples, worked out in detail, in the chapters that follow.

The question I want to ask here is, Can feminists hope to improve science by bringing into consciousness the implicit assumptions that underlie standard scientific descriptions and interpretations? I believe we can, which is why I think of myself as a scientist, even though I no longer do laboratory science. I consider myself a scientist because I continue to be keenly curious about nature and want to understand how it works. I do not think science as currently practiced is the way to find out, among other reasons because it proceeds by breaking nature into smaller and smaller bits, and then usually ignores, loses, or misreads their connections. Scientists engage in this practice, called reductionism, in the hope that once they get to understand the ultimate bits, they will be able to put them together and understand nature. But, as with Humpty Dumpty, that is going at it the wrong way around. I remain a scientist because I think we can understand nature. But we must be respectful of the connections and relationships within it.

Like most poets and novelists, scientists do not want their work to be read only as self-revelation and societal text. Writers are probing the psyche, human relationships, our place in society and the cosmos. Similarly, most scientists wish to probe nature, including people. In the current debate within feminist science criticism, I stand with those who argue that the political insights feminism provides can lead us to more accurate, hence truer, accounts of nature than we now have. But we must rigorously analyze the assumptions scientists make and specify as best we can the political, economic, and social realities that inform these assumptions. Maybe then we will be able to interpret scientific accounts—which are the stories scientists tell—in ways that allow us to understand nature less imperiously and to misuse and abuse it less.

PART ONE

How Do We Know?

The very idea that people can form a coherent, comprehensible picture of nature rests on assumptions not all cultures share. The British biologist and historian of science Joseph Needham (1969) has suggested that one reason what we call science arose in the West, despite the fact that at that time Chinese science offered a much clearer understanding of how nature works, is the Western assumption that nature is lawful and regular. This assumption has enabled scientists to formulate laws of nature analogous to the laws by which human societies operate. We say that Newton discovered the laws of motion, not that he invented them, as though these laws inhered in nature. We say that these laws permit certain events to take place and forbid others. We say that falling bodies obey the laws of gravity, and if a body rises into the air, we look for what made it do that, convinced that in a lawful universe it could not have done it on its own.

Needham argues that the fact that as late as 1730 a Swiss court prosecuted a rooster for laying an egg (and convicted him and burned him at the stake) suggests that the legal metaphor has been taken quite literally. He assumes, and I agree, that laws of nature came easily to the Western mind because of our belief in a humanlike god, who created the world and set it moving along a lawful, hence at least in principle a comprehensible and predictable, course. Needham writes: "The Laws of Nature which Kepler, Descartes, Boyle, and Newton believed that they were revealing to the human mind were edicts which had been issued by a supra-personal, supra-rational being." The very word *revealing* is symptomatic of our assumptions about science and nature. And although most people, and surely most scientists, no longer take the legal metaphor literally, it is built into the way we think about science and nature.

The way we make scientific facts and build them into coherent theories and descriptions sets limits to the kinds of things we can come to understand about nature. Scientists usually do not acknowledge these limits, nor do most other people. And the overestimation of science as a way to know, hence of the extent of the knowledge we can gain through science, has led us to undervalue other kinds of knowledge.

The scientific way to know has been labeled "objective" and identified as masculine; artistic, intuitive, and empathic ways of knowing are considered "subjective" and feminine. Thus knowledge has become gendered. And because the Western world-view values objectivity over subjectivity and men's knowledge over women's, "feminine" ways to know are by their nature inferior.

In this part of the book, I look at the conditions under which our science has developed and at the "subjectivity" of scientific fact making and knowledge. I point to the ways scientific work has been structured to accommodate the biographies of men, which makes it difficult for women to fit it into our lives.

These limitations are not inherent in the nature of women and men, any more than in the nature of science. They result from the ways scientific work, facts, and theories are constructed and from the ways we construct sex and gender, which I discuss in Part II.

1

Science in Context

EVER since the scientific method became a way of learning about nature, including ourselves, some people have hailed science as the only way to comprehend natural phenomena, while others have questioned whether it is an appropriate road to knowledge. As science and technology have grown, the questioning has deepened and expanded.

This is not to say that so-called scientific evidence is not still a good way to vouchsafe truth. Scientists' testimonies are used to endorse everything from toothpaste to nuclear power and weapons, but they are also used to challenge the very same things. And this is where the knife goes in because at present "scientific" support can be elicited on all sides of every question, so that the "lay" public is constantly forced to decide which scientists to believe.

Where then is the vaunted objectivity of science? People are realizing that they must either develop criteria on which to make these decisions (and to do so for each important issue) or decide to disbelieve all scientific explanations and look for other ways of knowing. (Incidentally, these other ways are sometimes no less empirical than the scientific ones.) The decision to disbelieve all scientific explanations is not to be sneered at. The volume, contradictoriness, and limited comprehensibility of much scientific information leave most people bewildered.

I am reminded of the comment Virginia Woolf attributes to her time-traveler, Orlando, who muses in 1928 as she enters the elevator at Marshall and Snelgrove's department store in London: "The very fabric of life now . . . is magic. In the eighteenth century, we knew how everything was done; but here I rise through the air; I listen to voices in America; I see men flying—but how it's done, I can't ever begin to wonder. So my belief in magic returns" (p. 196).

Not only "lay" people are ill at ease. After the defection of numerous physicists in the wake of Hiroshima and Nagasaki, the ranks of scientists closed and again became fairly solid. But this solidarity has begun to change, perhaps still as an aftermath of the questioning 1960s. Uneasy questions are being asked by scientists themselves. For example, the spring 1978 issue of *Daedalus,* the house organ of the American Academy of Arts and Sciences, is entitled "Limits of Scientific Inquiry." In the Introduction, Robert Morison, formerly director of medical and natural sciences at the Rockefeller Foundation, writes: "The scientific community has led a particularly unexamined life for a surprisingly long time, and many have accepted its unusual and, until recently, unquestioned status a little too easily. Indeed, in the last twenty-five years, in an effort to raise financial support at a rate nearly triple that of the rest of society, the scientific community may have promised too much too soon. Certainly it under-estimated the demand for accountability." And he goes on: "In all humility, it must . . . be admitted that it is impossible categorically to deny that we may have reached a point where we must abandon the faith that [in all cases] knowledge is better than ignorance. We simply lack the ability to make accurate predictions."

In the same issue of *Daedalus* Loren Graham, a historian then at Columbia University and now at M. I. T., begins his article with the sentence: "Concerns and anxieties about science and technology are not novel in the history of science, but a persuasive case can be made that a new stage has been reached in the last decade or so, a stage which has been marked not only by growing suspicion of science among segments of the lay public, but also by an expanding belief among scientists themselves that questions about the social responsibility of researchers must be faced more directly than has previously been the case." Although he focuses too narrowly on the issue of recombinant DNA and points to what he calls "graphic examples" of the "disastrous results" of "lay" interference in the directions of "fundamental research," Graham writes: "Society is deeply affected by science and technology in ambiguous and sometimes disturbing ways; some of the public concerns about what might happen if science and technology are left entirely to scientists and engineers are clearly legitimate."

The most revealing article in this symposium is by Robert Sinsheimer, chancellor of the University of California at Santa Cruz (and before that chair of the Biology Department at the California Institute of Technology). He suggests that it is imperative to "consider such issues as: the possible restriction of the rate of acquisition of scientific knowledge to an 'optimal' level relative to the social context into which it is brought; the selection of certain areas of scientific research as more or less appropriate for that social context; the relative priorities at a given time of the acquisition of scientific knowledge." Later on he writes:

> Institutions such as Caltech and others devote much energy and effort and talent to the advancement of science. We raise funds, we provide laboratories, we train students, and so on. In doing so we apply essentially only one criterion—that it be good science as science—that the work be imaginative, skillfully done, in the forefront of the field. Is that, as we approach the end of the twentieth century, enough? As social institutions, do Caltech and others have an obligation to be concerned about the likely *consequences* of the research they foster? And if so, how might they implement such a responsibility?

Answering his own questions, Sinsheimer writes:

> For reasons which probably need no elaboration Caltech has been more than reluctant to come to grips with this question. And, indeed, it just may be—and I say this with real sorrow—that scientists are simply not the people qualified to cope with such questions. . . . The focus must be on all the ties of the sciences to society and culture and on the impact of scientific knowledge and technological advancement on all human, indeed all planetary, life.

Although I have quoted from this symposium at some length, the level of its discussion is superficial. The basic assumptions of the practice of science are never questioned. There is no historical and political analysis of our hierarchically structured, exploitative society, in which scientists work not so much because they believe that the knowledge they produce is relevant to human needs or values but often in order to generate publications, jobs, research funds, and prizes. Because their analysis tends to be politically shallow, many scientists who have some social awareness and a sense of public responsibility flounder in

personal dissatisfactions that grow out of their inability to justify to themselves continuing to do scientific work once they
recognize how much of it involves personal aggrandizement at
public expense.

However, critics of science who ground their objections in the
counterculture or in one of the various non-Western ways to
knowledge often are ready to disavow science completely. This
position, I believe, throws out too much. Science offers important insights, but we must evaluate it critically from different
points of view. To make such an evaluation is the purpose of
this chapter.

Reification and the Language of Science

Scientific language helps to lend science an aura of depersonalized authority. One of its major hazards is reification, the
grammatical transformation by which a verb (in other words,
something someone does) is turned into a noun, a thing. Indeed the very concept, science, involves a reification because
there is no such entity as science. There are only the activities of
scientists. These activities are treated strangely in the language
of science because it is usual for that language to delete the
agent. When I report an observation, I do not write, "One sunny
Monday after a restful weekend, I came into the laboratory, set
up my experiment, and shortly noticed that. . . ." No; proper
style dictates, "It has been observed that. . . ." This sort of statement removes the relevance of time and place, and implies that
the observation did not originate in the head of a human observer, specifically in my head, but out there in the world. By
deleting the scientist-agent as well as her or his act of observing,
we are left with the scientific observation as a thing in itself,
which can be treated as though it were as real as nature and
comparable to it.

In the nineteenth century, when science was more of a cot-
tage industry, engaged in by "great men," these men sometimes
wrote in the first person singular. A rare instance of this style
that I remember seeing in more recent times is the opening
sentence of a scientific article by Jeffries Wyman (1965). I re-

member the joy with which I read: "In the course of reading the other day, at a window by the sea, the page proof of an article on linkage, I was suddenly struck by the realization that in all the years I had been thinking about the matter I had consistently failed to recognize one rather significant general concept." Never before or again has such a sentence appeared in the *Journal of Molecular Biology.* But immediately the charm is broken. The second sentence does not go on: "It is what I have decided to call. . . ." It reads: "This is what may be called. . . ." The thinking agent has disappeared; gone is the window by the sea; not another "I" in the paper; we are back in science. It was a momentary lapse, a minor declaration of personal independence.

Scientific writing implicitly denies the relevance of time, place, social context, authorship, and personal responsibility. Take as an example a sentence from E. O. Wilson's *Sociobiology: The New Synthesis,* a book published in 1975 that contains many such phrasings: "Human beings are absurdly easy to indoctrinate" (p. 562). Here the activity of one person, who has the power and desire to indoctrinate others, who presumably are not in positions to reciprocate, is turned into a pseudo-objective statement of fact—a fictitious description of "human nature."

But note that this seeming subtlety of grammar has important consequences for what logically follows. If our description of reality is "human beings are absurdly easy to indoctrinate," then it seems proper to ask what makes them be that way and why they might have evolved genes for "indoctrinability," which is what Wilson does. But if we describe the situation as one in which some people have the power to indoctrinate others, we might think about how to change the imbalance of power. The grammar of active participation is conducive to action, whereas the grammar of depersonalized description tricks us into submission to "facts of nature" or at best encourages their further exploration.

This kind of description, of course, is used widely in the social sciences. Take the concept of unemployment, which in the United States has become "chronic unemployment" or even "the normal rate of unemployment." By turning the activities of individuals who have the power to hire people, or not hire them, into depersonalized descriptions of fact, such phrases obscure a wealth of political and economic relationships that are subject to social action and change.

By turning activities of scientists into statements about nature or society, scientific language reifies them and in this way helps to mystify and intimidate the "lay public," those anonymous "others," as well as scientists. It leads them to accept situations as beyond human control and makes them feel powerless.

The Connection of Science to Nature

The impulse to understand nature rests in the belief (and it is only that) that the world is understandable, that it is orderly and operates on some logical plan that people can sort out. Scientific methodology is supposed to reveal the causal threads that enable us to explain natural phenomena.

Scientists can indeed construct threads of cause and effect, in which the effect is then interpreted as cause of a further effect, which in turn becomes cause, and so on. We can go further and weave such threads into the self-consistent fabric of theory that we call science. But we have not even begun to develop tools with which to tackle the question of what (if any) relationship that fabric bears to the seamless unity of nature out of which we have teased the original threads. The situation is even worse because we deduce both the correctness of our constructs and the essential identity of our fabric with the aspects of nature we try to comprehend from the fact that we can implement our understanding through action—by means of technology and other manipulations of nature. If we are wrong when we identify the threads or in the ways we weave them together, we risk damaging their original configuration in profound and perhaps irreversible ways that could threaten the integrity of the entire system, including ourselves.

I have used the image of threads and fabrics because they are ready metaphors for complicated, interweaving relationships. But they are misleading because they tempt us to think in terms of one- and two-dimensional systems. Yet in nature we have something intricate, patterned, seamlessly continuous, and many-dimensional that can be damaged or unraveled in many ways and perhaps beyond repair. The laws of entropy (among the most important, yet least widely understood, of nineteenth-century concepts) teach us that there are many different ways in which a complex, highly organized system can fall apart—which is to ascend the peaks of entropy—but that there is little or no likelihood that the separate parts will find their way back into their original entropy valley and arrange themselves as they were before.

There is a further difficulty and that is that our scientific methodology permits us to examine only those natural phenomena that are repeatable and measurable. It cannot deal with unique occurrences or with systems that flow so smoothly and gradually or are so profoundly interwoven in their complexities that they cannot be broken up into measurable units without losing or changing their fundamental nature. Furthermore, scientists tackle only the few, limited aspects that arouse sufficient interest or curiosity to engage private or governmental support. Our ways of doing science therefore elevate some things and events to the rank of "facts," indeed of "scientific facts," while we remain oblivious to the existence of others and relegate yet others to the foggy realms of supposition or, worse yet, superstition. The acceptance of science as the only proper way to learn about the natural world, including ourselves, locks us into seeing the world as the sum of those relationships and juxtapositions of objects and events that scientists can comprehend and analyze.

Whereas this scientific methodology obviously introduces elements of clarification and demystification by making it possible to describe and explain some aspects of nature, it inescapably distorts reality by redefining it on its own terms, while implying that they are the only appropriate terms. An element of mystification is introduced by the very exactness of scientific definitions and measurements. By virtue of their precise delineation, "facts" are removed from the context with which they are confluent. They gain an independent existence, which may be precisely the one that those who need them for their own purposes wish to confer upon them.

Our world of "facts" is contextual not only in that it depends on who we are and where and when, but also in that it is shaped by where we want our "facts" to take us. The Latin root of fact is *facere,* to make or do, and every fact has its *factor,* its maker. It exists as an autonomous unit only within the world in which its makers perceive a need for that particular "fact" because they want to build theories and actions on it or with it. But in the larger context these facts may be artefacts, and the theories and actions built upon them may disfigure or even destroy important aspects of nature.

If we become aware of these limitations and cautious in the ways we use the kind of knowledge to which the scientific method gives us access, we may be able to harness scientific insights to advantage. But we must realize that scientists' partial view and even more limited understanding may wreak havoc if we allow them to unbalance things too badly. We cannot afford to be Baconian warriors, who march fully armed into battle with nature. Notice that although Bacon urged people to "command nature in action," he warned that "Nature to be commanded must be obeyed." We are part of nature, and if we are not careful, its conquest may spell our defeat.

There is no need to be mystical about the effects of science. Science, as practiced, is flawed not because it is wrong to try to understand nature and use it but because too many scientists combine tunnel vision with arrogance, so that they constantly chew more than they can bite off.

Women as Objects Rather Than Makers of Science

Feminists are in a good position to analyze what is wrong with science as a social institution and with the knowledge it produces. We recognize that science was begun, and has been practiced, primarily by men from the educated upper classes in Europe and the United States and that it therefore embodies their values and views about how the natural world functions and should function. And although now Western science is also practiced in other parts of the globe, there, too, access is limited largely to economically privileged men who have been educated in European or U.S. universities or in local universities designed on the same model.

Although women have been the makers of very few "facts," they have been the objects of many scientific studies. Women and the other groups outside the mainstream that has generated the prevailing ideas and ideologies of the society must act like anthropologists, who observe the society and catalog its activities and the basic features of its explicit and implicit creeds. Nothing can be exempt because we realize that to define us as different (read "less") is one of the ideological and practical instruments that holds together the social order that oppresses us. To change our status as outsiders, therefore, we must see clearly how the social order functions to legitimate those on the inside. This insight is what consciousness raising is about—recognizing one's individual experience of oppression and understanding its significance within the wider setting that generates the oppression and defines its social function(s).

To succeed in our struggle for equality, women need to understand how scientific descriptions of ourselves as biological and social organisms are generated and used to maintain sexual inequality. Although women have not had a significant part in the making of science, science has had a significant part in the making of women. Science or, rather, scientists—that relatively small group of economically and socially privileged white men with the authority derived from being scientists—have had an important share in defining what women's human, and more specifically female, nature is and hence in defining what is normal for us to do and not do, indeed what we can do and be.

Even when we know deep down that most of what we have been told is nonsense—and again and again women have protested that it is—most of us do not possess the authority or the means to say so loudly enough to be heard and convince others. It is a fact of practical politics that the ruling group or class not only generates the reigning ideology but also controls the means that make this ideology the dominant or even exclusive "truth." This is so irrespective of political form—in democracies, where everyone can say what she or he thinks but is not able to publicize it equally, and in totalitarian states, where the discourse itself is limited.

As women, we have been not only scientifically misdefined but mistreated. Gynecologists and surgeons remove our female organs with a lack of concern that is in marked contrast to the respect they display for men's private parts. Indeed, as we will see in later chapters, a scientific/medical profession that has tried to exclude women from public life, lest our participation disturb the natural ebb and flow of our hormones, has not been equally hesitant to manipulate those very hormones—all, of course, in order to "help" us conceive or abort, or prevent conception or abortion, or keep from showing signs of advancing age.

Although we women should not prejudge what we will find when we critically examine the various attributes that supposedly characterize us, it is well to be suspicious when "objective science" confirms long-standing prejudices. When we hear scientific assurances that the domestic roles and tasks to which women have been relegated correspond to our innate biological propensities, we must ask whether the tasks have been adapted to women's biology or whether the notions of women's biology have been shaped to make us acquiesce in these tasks. It is always a mistake to accept too readily assignments of cause and effect for phenomena that occur together. Those very assignments may have been made to sustain the ideology that perpetuates their conjunction.

We need to understand that science is an abstraction from what scientists observe—necessarily a selective rendering of nature. And we need to emphasize that the abstractions and selections are made by people who live in a specific time and place and embody specific roles and interests. Although science is consensual, the imagination and thinking of scientists—that special group of people who share in the consensus—are no freer of contextual constraints than are those of other people. As long as the overwhelming majority of scientists are men who are rooted in the ruling class, socially or intellectually (or both), science will supply the "objective" supports and technical innovations needed to sustain patriarchal, ruling-class power.

But it would be naive to think that science under socialism is necessarily different. It is not. The roots of the problem lie in our Western view of nature, which leads us to seek power over nature rather than understanding from within it. Science developed in a specific social context. A rising merchant class needed to increase its understanding of nature in order to extend its control over it. When people want to use science in order to dominate and exploit nature for production, their scientific eyes see nature in exploitative ways. To use science to exploit human "others" is an easy extension of the subject/object relationship that underlies the scientific investigation and manipulation of animals and inanimate objects, as we shall see in Part III.

Use Value versus Exchange Value of Scientific Work

In his theory of production and value, Marx makes a crucial distinction between production for use and production for accumulation and exchange. This distinction is the origin of his concept of alienation—the discomfort that comes from laboring not for use and to satisfy needs but to accumulate unneeded, and often useless, products. Because capitalism depends on generating surplus value, capitalist production cannot be just for use; it must be for accumulation and exchange.

People need to exchange and accumulate goods, within limits, even near subsistence level. Above subsistence, stored goods can increase people's security by tiding them over times of scarcity. However, at present our level of production is far above what is needed for human comfort. The capitalist economy is geared to an ever-expanding accumulation of products, which are by definition useless because people's real needs have limits (although we are certainly not fulfilling everyone's needs). This situation is not only wasteful, but dangerous, because many of the products that are being "stockpiled" at ever-increasing rates and whose production and export keep the U.S. economy going are armaments of enormous destructiveness, so that our survival depends on their *dis*utility and *non*use.

Many people who agree with the broad outlines of this analysis do not realize that it is relevant also to the crisis brought on by the expansion of science as well as of technology. Most scientists believe that whereas technology is for production, science is "pure" and not for use. This is an artificial distinction. Since the beginning, Western science has been conceived as a way to harness nature for use. But it is true that much of the science that is produced at present is for accumulation and exchange, not for use. It is exchanged for fellowships, publications, jobs, research funds, positions on the committees that allocate funds, honors, prizes, and so forth.

When scientists work not because they want answers to the questions they ask of nature but for the exchange value the answers can bring on the job market (or the other "markets" I have mentioned), it becomes just as productive to discover something that is soon *dis*proved as to discover "eternal truths." The important thing is to realize the discovery's exchange value—in the form of publications, promotions, prizes—before someone else shows it is wrong. Indeed, in some ways, to discover something that is soon proved wrong has greater market value than to discover something with staying power because the "wrong" fact can earn reputations and jobs first for its discoverers, then for the people who correct it. A⁺ present, many scientists feel dissatisfied and alienated by this situation. They are disgruntled over the uselessness of their work and consider the exchange value they receive insufficient or of the wrong sort.

However, although science has developed historically as an adjunct to capitalism and has incorporated the capitalist ideology implicitly and usually without question, this ideology need not be part of science. Decisions about what science to do and how to do it can be made differently. The social context and use value of knowledge and its applications can be considered, and the power to make decisions about what to investigate can be shared among more and different kinds of people than are now involved in such decisions. If science were more useful and less alienating, mystifying, and remote from people's needs, the connections between scientific questions and our daily lives would be more obvious. In that case, more people could, and might choose to, get involved in making the relevant decisions.

Changing Science

It is pointless to argue for an end to scientific work. Something that can increase people's power so greatly cannot be eliminated by fiat. But science must be demystified and assessed within its political, social, and intellectual context. It must be seen as just one way to look at nature; a particular way to relate to organisms, including people; a very powerful, hence potentially dangerous, way to comprehend and use natural capacities and resources.

Social critics of science who are political activists must help develop an inquisitiveness about science among the people who are affected by its uses and thus stimulate this kind of evaluation. The scientific elite tries to present science as a value-free good that can be safely left in the hands of experts. In this way it helps to mystify science, but it is also a willing victim of this mystification.

True, the present contradictions of science will not be transcended without a radical restructuring of society, but it would be naive to assume that science will change automatically when the society changes. That will take an understanding of what is wrong with the way science is practiced and an active decision to change it.

2

Fact Making and Feminism

EVERY fact has a factor, a maker. The Brazilian educator Paulo Freire (1985) has pointed out that people who want to understand the role of politics in shaping education must "see the reasons behind facts" (p. 2). So the interesting question is, As we move through the world, how do we sort those aspects of it that we permit to become facts from those that we relegate to being fiction—untrue or imagined—and from those that, worse yet, we do not even notice so that they do not become facts, fiction, or figments of the imagination? In other words, what criteria and mechanisms of selection do scientists use to make facts?

One thing is clear: Making facts is a social enterprise. People cannot just go off by themselves and dream up facts. When they do that, and the rest of us do not accept the facts they offer us, we consider them schizophrenic, crazy. If their facts sufficiently resemble ours or they have the power to force us to accept them and make us see the emperor's new clothes, the new facts become part of our shared reality, and their making, part of the fact-making enterprise.

Making science is such an enterprise. As scientists, our job is to generate facts that help people understand nature. In doing this job, we must follow the rules of the scientific community and go about our fact making in professionally sanctioned ways. We must submit new facts to review by colleagues and be willing to share them with qualified strangers by writing and speaking about them (unless we work for companies with proprietary interests, in which case we still must share our facts, but only with particular people). If we follow proper procedure, our facts come to be accepted on faith by large numbers of people who are in no position to say why what we put out are facts, not fiction. After all, a lot of scientific facts are counterintuitive, such as that the earth moves around the sun or that if you drop a pound of feathers and a pound of rocks, they will fall at the same rate. Recently some physicists have even hypothesized that a pound of feathers falls more rapidly than a pound of rocks, a fact even more counterintuitive than what I learned in physics.

What are the social or group characteristics of people who are allowed to make scientific facts? Above all, they must have a particular kind of education, which includes graduate and postgraduate training. With such an education, they not only learn scientific subject matter but also must familiarize themselves with that narrow slice of history and culture that deals primarily with the experiences of Western European and North American upper-class men during the past century or two. In addition, they must not deviate too far from accepted rules of individual and social behavior, and they are socialized to talk and think in ways that let them earn the academic degrees required of a scientist.

Until the 1960s, youngsters mainly from the upper-middle and upper classes, most of them male and white, had access to that kind of education. Since then, more white women and people of color (both women and men) have been able to gain access, but the class origins of scientists have not changed appreciably. The scientific professions still draw their members overwhelmingly from the upper-middle and upper classes. How about other kinds of people? Have they no part in making science? Quite the contrary. In the ivory (that is, white) towers in which science gets made, one can find lots of people from the working and lower-middle classes, but they are technicians, secretaries, and clean-up personnel.

Decisions about who gets to be a faculty-level fact maker are made by professors, deans, and university presidents, who call on scientists from other, similar institutions to recommend candidates they think will conform to the prescribed standards. At the larger, systemic level, decisions are made by governmental and private funding agencies, which operate by what is called peer review: Small groups of people with similar personal and academic backgrounds decide whether a particular fact-making proposal has enough merit to be financed. Scientists who work in the same, or related, fields sit on each other's decision-making panels, and although criteria are supposedly objective and meritocratic, orthodoxy and conformity count for a lot. Someone whose ideas or personality or both are out of line is likely not to succeed.

Thus, science is made, by and large, by a self-perpetuating group: by the chosen for the chosen. The assumption is that if the science is "good," in a professional sense, it will also be good for society. But no one or no group is responsible for looking at whether it is. Public accountability is not built into the system.

What are the alternatives? How could we have a science that is open and accessible, a science for the people? And to what extent could—or should—it also be a science by the people? After all, divisions of labor are not necessarily bad. There is no reason and, in a complicated society like ours, no possibility that everyone be able to do everything. Destructive inequalities arise not from the fact that different people do different things but from the fact that different tasks are valued differently and confer different amounts of prestige and power.

For example, this society values mental labor more highly than manual labor. We think mental labor requires more specifically human qualities and is superior. This assumption is wrong and especially so in laboratory work because it means that the laboratory chief—the person "with the ideas"—gets the credit, whereas the laboratory workers—the people who work with their hands (as well as, often, their imaginations)—are the ones who perform the operations and make the observations that generate new hypotheses and permit hunches, ideas, and hypotheses to become facts.

But it is not only because of the way natural science is done that head and hand, mental and manual work, tend to be linked. Natural science requires a conjunction of head and hand because it is an understanding of nature for use. Natural science and technology are inextricable because we can judge our interpretations of nature only by seeing to what extent they work. Scientific facts and laws are true only to the extent that they can be applied and used.

Because feminist methodology is grounded in women's experiences and practices, it offers a lens through which to examine the ways scientific facts are related to the ideological and social commitments of scientists—the people who have been able to make scientific facts. By juxtaposing women's lived realities with the scientific descriptions of them, we can try to judge to what extent the "facts" explain our experiences or whether they misrepresent and conceal them.

Woman's Nature: Realities versus Scientific Myths

As I said before, to be believed scientific facts must fit the worldview of the times. For this reason, since the social upheavals that started in the 1960s, some researchers have tried to "prove" that differences in the political, social, and economic status of women and men, blacks and whites, or poor people and rich people are the inevitable outcomes of people's inborn qualities. They have produced "scientific" evidence that blacks are innately less intelligent than whites, and that women are innately weaker, more nurturing, and less good at math than men.

This kind of science is specious and ideology-laden. Clearly the ideology of woman's nature differs drastically from the realities of women's lives and indeed is antithetical to them. The ideology that labels women as the natural reproducers of the species and men as producers of goods ignores the fact that women produce a large proportion of our goods and services. But this ideology can be used to shunt women out of higher-paying jobs, the professions, and other kinds of work that require continuity and provide a measure of power over their own and, at times, other people's lives. Most women who work for pay do so as secretaries or nurses or in other jobs that involve a great deal of responsibility but do not pay well. For this reason, insisting on equal pay within job categories will not remedy women's economic disadvantage as long as women's jobs are less well paid than men's jobs. Scientists underwrite this stratification when they produce specious research that "proves" that girls are not as good as boys at spatial perception, abstract reasoning, mathematics, and science.

Discriminatory practices are often justified by the claim that they follow from limits biology places on women's capacity to work. In the 1970s, a number of women employees in the American chemical and automotive industries were forced to choose between working at relatively well-paying jobs that had previously been done by men or remaining fertile. In one instance, five women were required to submit to sterilization by hysterectomy in order to avoid being transferred from work in the lead-pigment department at the American Cyanamid plant in Willow Island, West Virginia, to janitorial work with considerably lower wages and benefits (Stellman and Henifin, 1982). Even though none of these women was pregnant or planning a pregnancy in the near future (indeed, the husband of one had had a vasectomy), they were considered "potentially pregnant" unless they could prove they were sterile. Although exposure to lead can damage sperm as well as eggs, it is as though only women can expect to be parents. In addition, lead exposure can affect the health of male and female workers as well as a "potential fetus."

It is important to notice that this vicious choice has been forced only on women who have recently entered relatively well-paid and traditionally male jobs. Women who are exposed to reproductive hazards in traditionally female jobs, as nurses, x-ray technologists, laboratory technicians, or as cleaning women in surgical operating rooms, scientific laboratories, or the chemical and biotechnology industries, or as beauticians, secretaries, workers in the ceramics industry, and domestic workers, are not warned about the chemical and physical hazards of their work to their health or to that of a fetus should they be pregnant. In other words, scientific knowledge about fetal susceptibility to noxious chemicals and radiation is used to keep women out of better-paid jobs from which they were previously excluded by discriminatory employment practices, even though, in general, working women and men are not protected against occupational hazards to health.

The ideology of women's nature that is invoked in such cases posits that a woman's capacity to become pregnant leaves her always physically disabled in comparison with men. The scientific underpinnings for these ideas were produced in the nineteenth century by the white, university-educated, mainly upper-class men who made up the bulk of the new professions of obstetrics and gynecology, biology, psychology, sociology, and anthropology. These professionals realized that they might lose the kinds of personal attention they were accustomed to getting from their mothers, wives, and sisters if women of their own class gained access to the professions. They therefore used theories about women's innate frailty to disqualify girls and women of their own race and class who might compete with them for education and professional status. But they did not invoke women's weakness to protest the long hours poor women worked in the homes and factories belonging to members of their own class nor to protest the labor black slave women were forced to do.

Nineteenth-century biologists and physicians claimed that women's brains were smaller than men's and that women's ovaries and uteruses required much energy and rest in order to function properly. They "proved" that young girls should be kept away from schools and colleges once they began to menstruate and warned that without this kind of care women's uteruses and ovaries would shrivel and the human race would die out. But, once again, this analysis was not carried over to poor women, who not only were required to work hard but often were said to reproduce too much. Indeed, scientists interpreted the fact that poor women could work hard and yet bear many children as a sign that they were more animallike and less highly evolved than upper-class women.

During the 1970s and 1980s, feminists uncovered this history. They analyzed the self-serving theories and documented the absurdity of the claims as well as their class and race biases and their glaringly political intent (Bleier, 1984; Fausto-Sterling, 1985; Hubbard and Lowe, 1979; Lowe and Hubbard, 1983).

But this kind of scientific mythmaking is not past history. Just as nineteenth-century medical men and biologists fought women's demands for equality by claiming that their reproductive organs made them unfit for anything but childbearing and child rearing, just as Freud declared women to be intrinsically less stable, intellectually inventive, and productive than men,

so, since the 1970s, has the renaissance in sex-differences research claimed to prove scientifically that women are innately
better than men at home care and mothering, while men are
innately better equipped than women for the competitive life of
the marketplace.

Questionable experimental results obtained with animals (primarily that prototypic human, the white laboratory rat) are
treated as though they can be applied equally to people. On this
basis, some scientists are now claiming that the secretion of
different amounts of so-called male hormones (androgens)
by male and female fetuses produces lifelong differences in
women's and men's brains. They not only claim that these (un-
proved) differences in fetal hormone levels exist but imply
(without evidence) that they predispose women and men, as
groups, to exhibit innate differences in ability to localize objects
in space, verbal and mathematical aptitude, aggressiveness and
competitiveness, nurturing ability, and so on (Goy and McEwen,
1980; Money and Ehrhardt, 1972; *Science*, 1981). Sociobiologists
claim that the differences in social behavior stereotypically said
to characterize women and men in Western societies (such as
male aggressiveness, competitiveness, and dominance, and female coyness, nurturance, and submissiveness) are human uni-
versals that have existed in all times and cultures. Because these
traits are said to be ever-present, sociobiologists deduce that
they must have evolved through Darwinian natural selection
until they became part of our genetic inheritance (Wilson, 1975).

In chapter 8 we will look in detail at how sociobiologists account for the differences in the positions of women and men
in society. For now, it is important to realize that as the new
scholarship on women has grown, a number of anthropologists
and biologists have been reevaluating the contributions female
animals make to social life and species survival. It is no longer
acceptable to think of females only in relation to reproduction
and to look to males for everything else (Hrdy, 1981, 1986;
Lancaster, 1975; Smuts, 1985). Scientists have challenged the
premises that underlie male-centered descriptions and analyses
of the female role and have offered thoroughgoing critiques and
new formulations (for examples, see Hubbard, 1979; Lewontin,
Rose, and Kamin, 1984; Lowe and Hubbard, 1979).

Subjectivity and Objectivity

I want to come back to Freire, who says: "Reality is never just simply the objective datum, the concrete fact, but is also people's [and I would say, certain people's] perception of it" (1985, p. 51). And he speaks of "the indispensable unity between subjectivity and objectivity in the act of knowing" (p. 51).

The recognition of this "indispensable unity" is what feminist methodology is about. It is especially necessary for a feminist methodology in science because the scientific method rests on a particular definition of objectivity that we feminists must call into question. Feminists and others who draw attention to the devices dominant groups use to deny other people access to power—be it political power or the power to make facts—have come to understand how that definition of objectivity functions in the processes of exclusion I discussed at the beginning of this chapter.

Natural scientists achieve their objectivity by looking at nature (including other people) in small chunks, which they treat as though they were isolated objects. And they usually do not acknowledge their own relationship to these "objects." In other words, natural scientists describe their activities as though they existed in a vacuum. We saw in chapter 1 how the way language is used in scientific writing reinforces this illusion because it implicitly denies the relevance of time, place, social context, authorship, and personal responsibility.

An example of the kind of context stripping that is commonly taken for objectivity is the way E. O. Wilson (1975) opens the last chapter of *Sociobiology: The New Synthesis.* He writes: "Let us now consider man in the free spirit of natural history, as though we were zoologists from another planet completing a catalog of social species on earth" (p. 547). This statement epitomizes the fallacy we need to get rid of. There is no "free spirit of natural history," only a set of descriptions put forward by mostly university-educated, Euro-American men who have had the necessary social and economic advantages to become scientists. Nor can we have any idea what "zoologists from another planet" would have to say about "man" (which I guess is supposed to mean "people") or about other "social species on earth," for how these "zoologists" were used to living on their own planet would determine how they would judge us.

Another example of the absurdity of pretended objectivity is a study reported in the *New York Times* in which scientists claimed they had identified eight characteristics in young children that could be used to predict the likelihood that they would develop schizophrenia later in life (Goleman, 1984). The scientists were proposing to estimate the risks for a group of children and then observe them as they grew up in order to test the accuracy of the predictions. Such experiments cannot be done. How do you find a control group for parents who have been told that their children exhibit five out of the eight characteristics or, worse yet, all eight characteristics said to predict schizophrenia? Do you tell some parents that this is so although it isn't? Do you not tell some parents whose children have been so identified? Even if psychiatrists agreed on the diagnosis of schizophrenia—which they do not—this kind of research would be highly unethical.

Feminists must insist on the dangers of denying our contexts and subjectivity. Although context stripping has worked reasonably well in classical physics, where falling bodies are treated as though they experience no friction, it must not be the model for every kind of science. In fact, practitioners of the new physics have recognized since the beginning of this century that the experimenter is part of the experiment and influences its outcome. That insight produced Heisenberg's uncertainty principle: the recognition that, in the realm of atoms and subatomic particles, the operations an experimenter performs disturb the system to such an extent that it becomes impossible to specify simultaneously the position and momentum of atoms and subatomic particles.

How about standing the situation on its head and using the social sciences, where context stripping is clearly impossible, as a model? How about always acknowledging the experimenter as a self-conscious subject who lives and does science within the context in which the phenomena she or he observes occur? We need to be like anthropologists, who take extensive field notes about a new culture as quickly as possible after they enter it, before they incorporate the perspective and expectations of that culture. They realize that once they know the foreign culture well and feel at home in it, they will begin to take some of its most significant aspects for granted and stop seeing them. But, at the same time, they must acknowledge the limitations their personal and societal backgrounds impose on the way they perceive the foreign society.

Awareness of subjectivity and context must be part of doing science because we cannot eliminate them. We come to the objects we study with our particular personal and social backgrounds and with inevitable interests. If we acknowledge them, we can try to understand the world, so to speak, from inside instead of pretending to be objective outsiders looking in.

The social structure of the laboratory in which scientists work, the community in which they live, and their interpersonal relationships are all part of the subjective reality and context of doing science. Yet we usually ignore them when we speak of scientists' work despite the fact that this work is done in highly organized social systems. The sociology of laboratory life is structured by class, sex, and race, as is the rest of society. We saw before that to understand what goes on in the laboratory we must ask who does what. What does the lab chief—the person whose name appears on the stationery or on the door—contribute? How are decisions made about what work gets done and in what order? What role do women, whatever their class and race, or men of color and men from working-class backgrounds play in this performance?

Other Ways to Do Science?

Women have actually played a large role in the production of science—as wives, sisters, secretaries, technicians, and students of "great men"—although not usually as accredited scientists. One of our jobs as feminists must be to acknowledge that role. If feminists are to make a difference in the ways science is done and understood, we must not try just to become scientists who work in the traditional institutions, follow established patterns of behavior, and accept prevailing systems of explanation. We must understand and describe accurately the roles women have played all along in the process of making science.

We must ask why certain ways of learning about nature and using that knowledge are acknowledged as science whereas others are not. I am talking of the distinction between the laboratory and that other, quite differently structured, place of discovery and fact making, the household, where women have developed different sorts of botany, chemistry, and hygiene as they have worked in gardens, kitchens, nurseries, and sickrooms. Much of this knowledge has been systematic, effective, and handed on by word of mouth and in writing. But just as our society downgrades manual labor, it downgrades knowledge that is produced in other than professional settings, however systematic it may be. It downgrades the orally transmitted knowledge, experiments, and observations made in households.

Here is a wide range of empirical knowledge that has gone unnoticed and unvalidated (in fact, devalued and invalidated) by the institutions that define what is to be called knowledge. Men's explorations of nature also began at home but later were institutionalized and professionalized. Women's explorations have stayed close to home, and their value has not been acknowledged.

What I am proposing differs from the project the domestic science movement undertook at the turn of the century. That movement tried to make women's domestic work more "scientific" in the traditional sense of the word (Newman, 1985). I am suggesting that we acknowledge the scientific value of the facts and knowledge women have accumulated and passed on at home and in the workplace.

I doubt that women as gendered beings have something new or different to contribute to science, but women as political beings do. One of the most important is to insist on the political content of science and on its political role. The pretense that science is objective, apolitical, and value-neutral is profoundly political because it obscures the political role science and technology play in underwriting the existing distribution of power in society. Science and technology always operate in somebody's interest and serve someone or some group of people. To the extent scientists claim to be neutral, they simply support the existing distribution of interests and power.

If we want to integrate feminist politics into science, we must also emphasize the political implications of the way science is done, taught, and otherwise communicated to the public. We must broaden the base of experience and knowledge on which scientists draw by insisting that a wider range of people must be able to do science and do it in a greater variety of ways and institutions. We must also insist that scientists provide information and understanding that can be useful to and used by different kinds of people. For this task, science would have to be different from the way it is now. The important questions would have to be generated by collective and consensual processes. A wide range of people would have to have access to making scientific facts and to understanding and using them. And research topics and facts that benefit only a small elite, while threatening to oppress and endanger large segments of the population, would be unacceptable.

The most concrete examples of a different kind of science that I can think of come from the women's health movement. For example, when the Boston Women's Healthbook Collective (1984) decided to revise their book and produce *The* New *Our Bodies, Ourselves* they called together a variety of women, representing a range of needs and interests, to set the agenda, as well as to do the necessary research and writing.

There is probably no single way in which to change present-day science, and there shouldn't be. After all, one of the problems with present-day science is that scientists narrowly circumscribe the allowed ways to learn about nature and reject deviations.

It is difficult for feminists who, as women, are just gaining a toehold in science to try to make fundamental changes in the ways scientists perceive science and do it. For this reason many of us scientists who are feminists live double lives and conform to the pretenses of an apolitical, value-free, meritocratic science in our working lives while we live our politics elsewhere. Meanwhile, many of us who want to integrate our politics with our work analyze and critique the standard science but no longer do it. Here again, feminist health centers and counseling groups come to mind as organizations that consciously try to integrate scientific inquiry with feminist political practices. Feminists who

are trying to reconceptualize reality and reorganize knowledge
and its uses in areas other than health need to create environments (out-stitutes?) in which we can work together and communicate with other individuals and groups who want to make
the scientific process accessible and equitable, so that people
with different backgrounds and agendas can trade questions,
expertise, and answers.

3

Women in Academia

Let us look in detail at some of the contradictions and problems feminists confront in academia. After all, the universities and their academic tradition are archetypal male institutions, if anything even more male than their clerical forerunners. Nuns and abbesses once held a great deal more power than most female academics do. Even Mary Ritter Beard, who believed that women played a much more significant part in Western history than either traditional or feminist historians usually grant, spoke of the academic curriculum as "basically a sex education—masculine in design and spirit. Its tissue consists of threads instinctively selected from men's activities . . . woven together according to a pattern of male prowess and power as conceived in the mind of man. If the woman's culture comes into this pattern in any way, it is only as a blurring of the major concept" (in Lane, 1977, pp. 206–207).

Feminist scholars recognize the limited perspective of standard scholarship in literature, history, anthropology, sociology, and psychology. But the realization that the accepted view of ourselves as biological and social organisms is part of the same exercise in "sex education" comes as a shock even to many feminists and especially to feminist natural scientists. Yet it is crucial that we recognize that our view of the way human organisms have evolved and now walk, talk, work, and procreate is a male view in which the female is a variant on "mankind's" basic plan.

One reason for the dominance of this male view is that the outlines of our present-day biological self-concept were sketched in the nineteenth century, which produced the natural and social scientists and philosophers who generated most of our present political and scientific ideologies. God's laws could no longer be called on to justify inequality. Yet the revolutionary doctrines of liberty, equality, and *fraternity* and of the rights of *man* were not intended to apply to women. Nor were they meant to level the developing class structure of the growing industrial capitalism that spawned these movements for equality. Therefore the new social and scientific thinking had to explain why and in what ways women, poor people, immigrants, slaves, and colonial subjects—indeed all the disenfranchised or unenfranchised—were by nature unfit to profit from the opportunities offered by the bourgeois revolution. One of the socially most significant accomplishments of nineteenth-century science was to replace the waning ideological power of the church by bolstering or replacing God's laws with laws of nature.

I am not suggesting that scholars and scientists consciously conspired to produce fraudulent results to legitimate the beliefs and actions of the ruling elite. I am saying something more interesting: The fact that they were male and almost exclusively scions of the ruling classes led them to interpret the world so as to explain, with minimal risk to their positions in society, why they and their like were entitled to education and power, whereas other kinds of people were not.

Because their power was under attack by "shrill" feminists and their male supporters, such as John Stuart Mill, and traditional ideological means of stilling such protests were being eroded by the egalitarian spirit of the age, it is not surprising that Victorian gentlemen like Darwin, Spencer, Huxley, and Galton should describe biological and social relationships in ways that would uphold their accustomed power and privilege.

Arguments for Excluding Women from the Academy

Nineteenth-century scientists "proved" that women's brains were inferior in size and capacity (as were those of American Indians and blacks), but this kind of research was relatively easy to dismiss. As girls gained access to the same education as boys, it became clear that their brains were up to the task. Women's colleges attracted successful students and so did the adventurous coeducational colleges and universities.

By the 1860s and 1870s scientific arguments about the inferiority of women's brains were no longer acceptable in most quarters. The question was no longer, Can women do it? New reasons were required to limit women's access to higher education and the professions. So the question became, Is it good for them? And who would answer this question better than physicians, scientifically trained men who had been caring for the health of women?

The times were problematical for "regular" physicians, the men who had been able to get medical training in Europe or in the few American medical schools that were beginning to model themselves on the European schools. Women were prominent in "irregular" healing, where husband/wife teams were not unusual. Most midwives were women, and women had always cared for the ill at home. So, when women began to apply to the medical schools where the "regulars" were being trained, male physicians had reason to worry that becoming a physician might look like a natural extension of what women had been doing all along. What is more, the prevailing modesty of upper-class women was likely to make many of them prefer to be treated by a woman physician if they had the choice, and this preference was likely to have their husbands' approval. Physicians therefore viewed the entry of women into the medical profession as an economic threat, particularly at a time when they had begun to think the profession was already overpopulated. It was convenient for them to believe, and explain, that it would be detrimental to women's health to be educated the same as men.

In 1869 Dr. Edward H. Clarke, who held the chair of Materia Medica at Harvard Medical School, published an article in the *Boston Medical and Surgical Journal* in which he argued that women had a right "to every function and opportunity which our planet offers, that man has," if they are equal to it. But he advised moderation: "Let the experiment . . . be fairly made . . . [and] in 50 years we shall get the answer" (quoted in Walsh, 1977, pp. 120–121). However, Clarke soon turned against the notion that women should be able to attend medical school together with men. Let women be educated to the best of their capabilities, he urged, but separately.

Even if women had wanted to take Clarke up on his experiment, they could not follow his advice because medical schools were operated by and for men. But they kept applying, and even to Clarke's own Harvard Medical School. Under this threat, Clarke drew back from his proposal and in an address to the New England Women's Club of Boston, delivered in 1872, put forward the arguments he published in 1874 in his book *Sex in Education; or, A Fair Chance for the Girls.* In the next thirteen years Clarke's book went through seventeen editions, and a bookseller, who claimed to have sold two hundred copies in one day, remarked that "the book bids fair to nip coeducation in the bud" (quoted in Walsh, 1977, p. 124).

In the introduction, Clarke emphasizes that the point is not that one sex is superior to the other but that the sexes are different, and for this reason coeducation is detrimental to women's health. A girl obviously can go to school and do whatever a boy can do. "But it is not true that she can do all this, and retain uninjured health and a future secure from neuralgia, uterine disease, hysteria, and other derangements of the nervous system, if she follows the same method that boys are trained in. . . . The physiological motto is, Educate a man for manhood, a woman for womanhood, both for humanity. In this lies the hope of the race" (p. 16).

The rest of the book is built around this physiological rallying cry. In the chapter entitled "Chiefly Physiological," Clarke describes in painful detail all the ways women's physiology would be damaged if girls were educated the same as boys and therefore developed their brains rather than paying due regard to their menstrual functions. "The organization of the male grows steadily, gradually, and equally, from birth to maturity," he states authoritatively, whereas girls pass through a critical transition when they must "allow a sufficient opportunity for the establishment of their periodical functions. . . . Moreover, unless the work is accomplished at that period, unless the reproductive mechanism is built and put in good working order at that time, it is never perfectly accomplished afterwards" (p. 38). And it is not enough for girls to be careful how they *begin* to menstruate. Clarke cites "the accomplished London physician, and lecturer on diseases of women," Dr. Charles West, to the effect that "precautions should be . . . repeated . . . again and again, until at length the *habit* of regular, healthy menstruation is established" (p. 39). Clarke continues: "There have been instances, and I have seen such, of females in whom the special mechanism we are speaking of remained germinal, undeveloped. . . . They graduated from school or college excellent scholars, but with undeveloped ovaries. Later they married, and were sterile" (p. 39).

Here Clarke was speaking to a fear that was prevalent among the native-born upper classes. The cry had gone up that they were not having enough children while poor, immigrant women were having too many. Therefore the threat that education would render upper-class women sterile was something to alarm both women and men of his class.

I have quoted Clarke at such length because it is important to see the ways he transformed some of the prejudices of his time and class into scientific truths. He did not go unchallenged. For example, Thomas Wentworth Higginson (Emily Dickinson's friend and sponsor) pointed out that Clarke's dire descriptions meant little as long as he offered no comparable information about how education affected the health of boys (Walsh, 1977). Other critics objected that his proposal that girls' education be coordinated with the demands of their menstrual cycles would be equivalent to no school for girls because female pupils do not menstruate in synchrony. And what of the teacher, wondered Eliza B. Duffey: "She too requires her regular furlough, and then what are the scholars to do?" (quoted in Walsh, 1977, p. 128). Duffey further asked whether Clarke wished to suggest that wives and mothers require three- or four-day menstrual furloughs each month.

Despite Clarke's best efforts women gradually gained access to higher education, but his warning was worded strongly enough to worry those who wanted to educate women. "We did not know when we began whether women's health could stand the strains of education," wrote Martha Carey Thomas, first president of Bryn Mawr College. "We were haunted in those days by the clanging chains of that gloomy little specter, Dr. Edward H. Clarke's *Sex in Education*" (quoted in Walsh, 1979, p. 115).

Another pointed warning was sounded in an address to his fellow members by one Dr. Withers Moore, president of the British Medical Association, who proclaimed that to allow women access to higher education "will hinder those who would have been the best mothers from being mothers at all, or, if it does not hinder them, more or less it will spoil them, and no training will enable themselves to do what their sons might have done. . . . Unsexed it might be wrong to call her, but she will be more or less sexless. . . . And the human race will have lost those who should have been her sons. Bacon, for want of a mother, will not be born. . . . Women are made and meant to be, not men, but mothers of men" (British Medical Association, 1886).

Many of the roots of our biological self-concepts lie in this period. And what is so nice about quoting nineteenth-century "experts" is that the ideological basis of their pronouncements is obvious, whereas today prejudices tend to be voiced more discreetly.

Don't misunderstand me: I am not reciting a litany of disadvantage and self-pity. I want to emphasize not only that our history and psychology are male constructs, but that our biology also was not created by God the father but by his scientific and medical sons. When they studied women's biology and noted quite accurately that there are ways in which it differs from their own, they interpreted these differences as reasons to disqualify women on scientific grounds from participating in their world.

More than that, they judged women's physiology as not just different but abnormal, and classified women's life cycle into five debilities: menstruation, pregnancy, childbirth, lactation, and menopause; to which more recently have been added the disabilities of fecundity as well as of its absence, infertility. Two medical specialties, obstetrics and gynecology, were constituted, and in the name of cures for women's "diseases" came routine interference with women's normal physiology: in the nineteenth century, medicalized childbirth and gynecological surgery at a time when surgery regularly resulted in fatal infections, and in our day medical interventions in every aspect of procreation— fecundity, infertility, pregnancy, childbirth, and lactation—as well as menstruation and menopause.

I will say more about medical interventions and the reasons for them in later chapters. For now, I want to repeat something I heard Wilma Scott Heide, former president of the National

Organization for Women, say at a meeting on feminism in academia: "All knowledge is suspect." She was not being anti-intellectual. On the contrary, her point was that once we understand that most of what we "know" has been certified by a male elite that defines female experience as irrelevant to the mainstream of human thought and action, we need to pick up each piece of knowledge and look it over carefully to see how it fits into the whole picture. Nor need we accept the ways knowledge has been fragmented to create bits of expertise that do not enrich us but in fact decrease our understanding and autonomy.

An Agenda for Feminist Academics

The agenda for feminist scholars is thus vast. Among other things, we need to reevaluate the divisions of our daily lives into productive versus reproductive labor, into public versus private spheres, into work versus leisure: divisions that make little or no sense in the lives of most women, nor indeed of many men. Furthermore, we must inquire into how the knowledge we generate is spread: who has access to it and who makes decisions regarding its use.

These are some of the questions that are usually deemed irrelevant in traditional research, precisely because the answers direct our attention to the unequal distribution of power and privilege that underlies scholarly and academic establishments as much as business and enterpreneurial ones.

Women have other problems in the academy. Whether because our lives are punctuated by physiological transitions that make individual variations in experiences obvious or because we are used to living with children, whose lives unfold organically, or perhaps for other reasons, the rigid, hierarchical structure of academic life, with its artificial rites of passage that must occur in the proper sequence and at the proper times, is particularly hard on us.

As I discussed in the first chapter, during the past hundred years scholarly labor has moved away from the feudal model and has become alienated, like the rest of work under industrial capitalism. The most valued reward of our achievements is not our intellectual and aesthetic satisfaction with our work, but the money, power, and prestige it brings. Our work cannot be guided by internal rhythms or by a dynamic interplay of curiosity, insight, inspiration, and utility. It must produce marketable results that will bring publications, jobs, promotions, prizes. And thus the questions we ask must be readily answerable, hence narrow in their focus and definition.

Some of these problems exist for all academics who come to scholarly and scientific work with the illusion that university life will nurture their productivity, unencumbered by market pressures. But the capitalist mode of production in the academy is particularly hard to integrate into women's lives because it generates a lot of stresses during the time when we must bear children, if we decide to have biological offspring. The synchrony of market demands with the demands of our procreative lives, which involve unpredictable personal relationships and body rhythms, may be among the stresses that led feminists like Shulamith Firestone (1972) to argue for technological procreation in "plastic wombs."

Problems also arise if we try to evaluate every fact to see whether it is "sex education." After all, power over the quality and content of our jobs rests in the hands of men who may feel threatened by women, even when we accept the traditional definitions of what is real and true. As Virginia Woolf ([1931] 1977) said, men "had thought that nature had meant women to be wives, mothers, housemaids, parlormaids and cooks. Suddenly [they] discover that nature . . . had made them also doctors, civil servants, meteorologists, dental surgeons, librarians, . . . and so on."

The disaster is compounded when feminists come up with research questions and results that transgress the boundaries of intellectual propriety or, worse, threaten established power relationships. Returning to Woolf: It is not pleasant for a man who "has been out all day in the city earning his living, and he comes home at night expecting repose and comfort, to find that his servants—the women servants—have taken possession of the house." It is even worse if they have started to put out the rugs and rearrange the furniture, or—heaven help us—have put the house up for sale.

4

Women Scientists

The issue of gender gets into the social analysis of science because of the time and place in which modern science developed. There is no intrinsic reason why most scientists should be men, but during the past two or three centuries, as scientific knowledge and its applications have expanded into a dominant world-view, men have become so identified with all phases of science—the activity, product, application, and world outlook—that it has come to be thought of as essentially masculine. In 1982, Margaret Rossiter published *Women Scientists in America: Struggles and Strategies to 1940,* which documents the steady, careful, and politically astute efforts women made between approximately 1820 and 1940 to join the scientific world and the different barriers male scientists erected to make their entry difficult. The topic and Rossiter's handling of it are sufficiently important to the argument I am developing in this part of the book that we need to take a look at it.

Even some feminists postulate a quasi-natural antipathy between women and science. For example, Evelyn Fox Keller (1985) has suggested that because of the early psychosocial experience of being nurtured by women, men early on learn to objectify and manipulate people and the environment, whereas women learn to identify with others and to be comfortable with relationships involving mutuality and nurturance. True, because science has been produced largely by men, it bears marks of its male parentage, such as objectivization and instrumentality. But I see no reason to believe that women and men divide into nonoverlapping categories with respect to instrumentality and domination of nature any more than we do in our other, not directly procreative, characteristics. The exclusion of women from science is too dramatic and quixotic to be explained on the basis of differences in intellectual or temperamental preferences between women and men as groups.

For example, Rossiter has shown that women were far more successful in botany than zoology, in astronomy than physics, in statistics than mathematics, and in anthropology than geology or paleontology. One must look to the history and growth of professionalization in the various areas of science rather than to individual or group psychology for explanations of such strange occurrences. Why women were better represented in nutrition, with its links to home economics, than in chemistry, or in psychology than in biology, can perhaps be explained by the congruence of the supposed differences among these subjects with the ideology of "woman's nature"—her supposedly innate preference for domesticity and human relationships. But the other pairs lack that easy social logic and can be explained only by what else was happening at the time these fields of inquiry came into prominence or were professionalized; by the political and economic forces groups of well-placed women could muster; or by their ability to find effective male allies who were not biased against women, were themselves feminists, or preferred to surround themselves with women for sexist reasons.

The fact that women were more successful in gaining access to biochemistry than to anatomy and physiology has its parallel in the fact that biochemistry was also more accessible to Jews than were anatomy and physiology. It may be that the young, aspiring science of biochemistry was simply more open than the old, established medical sciences.

Yet, overall, women's early hope that if they became educated, they would be accepted as the equals of similarly educated men was not fulfilled. To be permitted to get an education, upper-class women had to compromise and agree that they would become educated so as to be better wives and mothers, not to practice what they were learning in order to earn their living, the same as men. Later, when college-educated women began to encroach on men's professions, the men with power erected new barriers, such as requirements for advanced degrees, which women could not get. Once women, too, began to earn Ph.D.s, universities proved far readier to accept them as students than to employ them as professors.

It was well-nigh impossible for women, even if they had doctorates, to get the kinds of jobs equally or less qualified men were getting. Women had to take low-paying, dead-end jobs assisting men in university, government, or industrial laboratories, or work in stereotypically feminine disciplines, such as home economics or, later, child psychology. The only institutions that made a modicum of room for them were the women's colleges. But even they sought the prestige men brought. They also tended to limit the amount of research their science professors could do, whether these were women or men. The difference was that faculty women had fewer opportunities than men to move on to professorships in prestigious universities. Numerous well-known male scientists began their academic careers at women's colleges and later moved on to places like Harvard, Columbia, or the University of Chicago, where they could get the stimulation and support they needed to advance their research. When women began to apply for membership in scientific societies, new barriers were erected there as well, such as two-tiered "associate" and "full" memberships, all in the name of "raising standards."

During the most active period of the first wave of feminist politics, between about 1890 and the early 1920s, women collected funds to open opportunities in coeducational institutions, such as Johns Hopkins. And they endowed research spaces specifically reserved for women at international laboratories, such as the Zoological Station at Naples, Italy, and the Marine Biological Laboratory in Woods Hole, Massachusetts.

Women also began to prepare reports to document discrimination in employment. However, without legal sanctions women were entirely dependent on the goodwill of individual men and nothing much changed. The patterns of sex-segregated employment were so thoroughly established in academia, government, and industry that by the 1920s, when numbers of women had earned doctorates in the sciences, neither the ability to prove discrimination nor the weak kinds of political pressure women could muster produced jobs and recognition, except in rare, individual cases.

The resulting tokenism, of course, was to the disadvantage of the majority of women. By celebrating, and occasionally hiring, an "exceptional" woman, men could profess fairness while continuing to pass over women who perhaps were no better, but also no worse, than the men who were being hired. Women's brave accommodation to this cruel reality and their resignation to having to achieve more for fewer rewards is a sad story. And it is infuriating to see how explicit and unashamed the discrimination has been. Rossiter cites the Nobel physicist Robert Millikan's advice in 1936 to the president of Duke University not to appoint the distinguished physicist Hertha Sponer, a refugee from Nazi Germany, because chances of building an outstanding department would be better if he "picked one or two of the most outstanding younger men, rather than . . . filled one of [the] openings with a woman" (p. 192).

The role of the women's colleges as producers and employers of women scientists is interesting. As a Radcliffe A.B. and Ph.D., I have always thought that Radcliffe struck a bad bargain for women when it decided to contract with Harvard for its professors rather than build a faculty of its own. It meant that as women scholars and scientists became available, Radcliffe was not able to hire them. It also meant that when Radcliffe boasted that its students had the privilege to study under Harvard's "great men," it was ignoring the fact that it was thereby denying us the realistic expectation that we might some day be equally great women. Indeed, the only examples of great women Radcliffe could provide for its students were its administrators (although during part of my days as a student even Radcliffe's president was a man) and the faculty wives who "helped" their husbands become "great men" by working with them as unpaid and usually unacknowledged collaborators or by serving as hostesses at academic social events. Small wonder Rossiter found that fewer women listed in the 1938 edition of *American Men of Science* earned their undergraduate or graduate degrees at Radcliffe than at any of the other "seven sisters" and at many coeducational colleges.

Again, the point is not to complain at sexist injustices, but to recognize them for what they are and root them out. In the next chapter I want to elaborate on the well-known story of Rosalind Franklin, a scientist who made major contributions to the birth of the now-flourishing field of molecular genetics (often called molecular biology), but who was erased from the record for more than two decades until Anne Sayre (1975) and the women's movement forced the reevaluation and acknowledgment of her work.

5

The Double Helix: A Study of Science in Context

I want to end this part of the book with a detailed look at how the double-helical model of DNA came into being. The importance of DNA in modern biology and medicine might be reason enough to examine how its structure was established. But I want us to look at it because it offers a striking example of how personal and societal agendas, commonly thought to lie outside the realm of objective science, are part of the process of scientific discovery.

Autobiographical

I came to the United States after the Nazis took over Austria in 1938 and attended high school and college in pre-World War II and wartime America. Ten years later, I recrossed the Atlantic and spent the year working in a laboratory in London. It was the postwar London of rationing, queues, and bomb sites, full of the elation of victory over German bombs and missiles (familiarly known as doodlebugs), with a Labour government, elected by a strong majority, that was introducing a national health service soon to become a model for the rest of the world. It was a drab and dowdy London, drizzly and foggy and cold; a happy and optimistic London, full of well-fed and well cared-for children, Britain's special wartime concern and pride; the London I fell in love with and the London a young English chemist named Rosalind Franklin gladly exchanged for Paris—at least for a time.

The following year, I came back to the United States to finish
my Ph.D. But I was back in Europe the summer of 1950, at the
Carlsberg Laboratory in Copenhagen, where the Carlsberg
brewery rationed us to no more than six bottles of beer (or soda)
a day. When I visited the laboratory of the Danish biochemist
Hermann Kalckar, I found him all excited because two young
scientists were about to arrive from Max Delbrück's laboratory
at the California Institute of Technology: Gunther Stent and
Jim Watson. Watson (1968) has written about his Copenhagen
period in *The Double Helix.*

When I returned to Copenhagen two years later, to work
there for a year, Watson had moved on to Cambridge, England.
He hadn't accomplished much in Copenhagen other than ac-
company Kalckar to Naples, where Maurice Wilkins had shown
him an x-ray photograph of DNA. The world found out what
he was doing in Cambridge in April 1953, when Watson and
Francis Crick published their instantly famous paper in *Nature*
in which they proposed the double-helical model of DNA.

Five weeks later, on that rainy day in June 1953 when the
London streets were packed with people come to see Elizabeth
II crowned Queen of England, I found myself quite by accident
in the Cavendish Laboratory in Cambridge, listening to Crick
expound with great excitement on his and Watson's newly discovered model of DNA. I did not meet Watson until he came to
Harvard a year or two later. We became friends and remained
friends until disagreements about Rosalind Franklin and the
controversy surrounding the uses of recombinant DNA technology and genetic engineering stretched our relationship beyond the breaking point.

My years of scientific apprenticeship and early intercontinen-
tal wandering overlapped Watson's. We spent our scientific adolescence during the same period in similar places. When I read
The Double Helix, I recognized attitudes, friends, and sights.
And I liked the irreverence with which Watson described scientists and science. I had not heard of Rosalind Franklin or met
her, and such was my state of integration into male-dominated
science that I noticed nothing wrong. (Let me say at once that
many others, who also did not know Franklin, were a good deal

more alert than I to Watson's outrageously sexist treatment of her as a woman and a scientist.)

I first became aware of the priority scandal surrounding the Watson-Crick model of the double helix when a friend, herself an x-ray crystallographer, loaned me the manuscript of Anne Sayre's (1975) book about Franklin. I have since confirmed many of Sayre's facts and spoken with her and a number of Franklin's other friends as well as read what the main actors and several commentators have written about this episode. It is on this basis that I have formulated my appraisal of what took place.

Philosophical

The art historian Ernst Gombrich (1963) has quoted one of his colleagues as saying that "all pictures owe more to other pictures than they do to nature" (p. 9). And so it is with science. Of course, nature is in there; but the primary link is with the science that has gone before. In a way that is only to say that, like art, science is a product of the human imagination, and, like art, to be acceptable and have meaning, it must be comprehensible in terms of our experiences of the world and of the accepted interpretations of it. When we make science, we abstract from the seamless unity of nature the things we notice and therefore pull out of the continuum. Alan Watts ([1958] 1970, p. 55) says: "Things are the measuring units of thought just as pounds are the measuring units of weighing." And just as the number of pounds in a rock or of inches in a log offers only a one-dimensional description of these objects, so the "things" or "facts" we use to describe our experience of nature are only discrete fragments of its total reality. When we translate these units of thought and experience into words, we trim them even further because, as Watts points out, "the fundamental realities are the relations . . . in which facts [or things] are the terms or limits" (p. 57). So, if the world consists of an infinitely intertwining web of relationships, which our sensations and thought processes sort into things and facts, and which we delimit still further when we name them, science raises these stages of abstraction (or, more correctly, of concretization or reification) yet another notch. Scientists make a further selection when they turn facts of nature into facts of science.

Scientists do not just hold up a mirror to nature. They use something more like a coarse sieve through which fall all the things they don't notice or take to be irrelevant. The intellectual labor of scientists consists of constructing a coherent picture of the world from what they sift out as noteworthy and significant. Anthropologists use the terms *backgrounding* and *foregrounding* to describe what they do when they enter a foreign culture and try to understand the rules by which it operates. These words denote the often unconscious activity by which all of us allow certain things, habits, and events to merge into the background of the unnoticed or "unnoteworthy," while we pull others into the foreground, where they command attention and have to be dealt with. A similar process occurs in the natural sciences when scientists construct juxtapositions and relationships out of the "facts" into which they sort the complex web of the real world. To alter Watts's image, we can say that scientists transform the seamless unity of nature into a carefully patterned patchwork quilt. And we need to be aware of the elements of both patchwork and patterning, for both involve choices that are far from arbitrary.

The way the patches are selected—the unconscious decisions of what remains background and what gets pulled into the foreground—and the way they are stitched together are determined only in part by the explicit postulates and rules of the scientific method because they, like our selective mind-set, are social products that depend on who we are, where, and when. Our scientific reality, like all reality, is a social construct. And by that I do not mean that there is no real world out there, only that what we see and how we interpret it depend on the larger social context.

We learn to see the world while we are infants and young children by a process that is guided by all the people around us. By the time we are educated (literally, "led out"), we cannot remember how we saw the world before we learned what it was "really" like. Learning to speak in part involves precisely this kind of acquisition of the important social, hence verbal, categories and distinctions. When a child has been corrected often enough that "this is a doggie not a kitty," the child has learned not only two words but also that there are different kinds of domestic animals with fur and a tail, and that it matters to know them apart. And when a child learns much subtler distinctions, such as "no, dear, this is a boy, not a girl," the very subtlety, coupled with the insistence that the distinction be made correctly every time, conveys a profound social message.

In this way, we internalize the basic rules and concepts of our society quite without knowing it. Usually only people who are outsiders to the prevailing belief system become sufficiently conscious of the learning process to question its results. That is what consciousness raising is about.

Sociological

The aspects of the DNA story on which I want to focus have to do with styles of work in science, with what one considers important in it, and with the ways women and men live in scientific laboratories.

The first question I want to ask is, Why did Watson (1968) call DNA "the most golden of all molecules"? And why did he, a bright and ambitious nineteen-year-old, former Quiz Kid, decide that it was the key to understanding life as well as, perhaps not incidentally, to his own success?

Perhaps in a less individualistic society that cared less about the traits, accomplishments, and successes of individuals, people might care less about how such traits pass from parents to their children. In such a society, people might find many aspects of biology more interesting than heredity, genes, and hence (in a reductionist science) DNA. Yet, as I have said before, beginning with the period of the American and French revolutions, the meritocratic philosophy of Euro-American liberalism, combined with the obvious unwillingness of the ruling classes to give up power, made it mandatory that the men from the educated, upper classes find biological reasons why they were entitled to rule the world. This impulse was later carried to caricaturish extremes by the Nazis, who made pronouncements such as

"National Socialism is nothing but applied biology" (quoted in Lifton, 1986, p. 31).

But even in a highly competitive, individualistic society scientists might not expect to answer the question What is life? by studying genes if they did not accept Descartes' definition of the organism as a machine. This definition leads biologists to expect to find out what life is by taking living organisms apart and reducing them to smaller and smaller units—eventually chemistry and physics—even though the attributes we associate with being alive are lost in the process. Nor might biologists expect to find out how genes work by analyzing DNA, as is done now in what Watson and Crick, with characteristic modesty, named molecular biology, thus implying that DNA is the only molecule that matters in biology.

I emphasize this reductionism because Watson has written that, long before they knew each other, he and Crick acquired their keen interest in genes and DNA by reading Erwin Schrödinger's (1944) *What Is Life?* Written by one of the great physicists, this little book drew the attention of physicists to biology at the end of World War II, when many of them were becoming disillusioned with physics. After all, by 1945 the proud physics of relativity and quantum theory and of the principles of complementarity and uncertainty had generated the atomic bombs that destroyed Hiroshima and Nagasaki. Intellectually as well, physics was beginning to degenerate into queuing up in front of bigger and bigger machines so as to produce smaller and smaller particles. Many physicists had begun to look for more interesting problems and responded to Schrödinger's challenge of the gene as the new frontier.

In the story of the double helix, the only main actor trained as a biologist was Jim Watson. The others were physicists or chemists. Moreover, Delbrück, on whose ideas Schrödinger based his discussion of the gene, was promising not only that life would ultimately be understood by reducing it to physics (Delbrück did not like chemistry), but that the physical exploration of life would lead scientists to discover new physical laws. This was one reason Delbrück, and perhaps some other physicists, shifted to biology. Watson (1966) has written that one of the reasons he opted to get his Ph.D. with Salvador Luria was because Luria and Delbrück had worked together and were friends. And, indeed, his association with Luria soon brought Watson under Delbrück's wing.

In the event, understanding the structure of DNA has produced no new physical laws nor has it taught us what life "is," although we have learned a lot about the chemistry and structure of proteins and nucleic acids, those big molecules that occur by and large only in living organisms. Nor has knowing the structure of DNA revolutionized our thinking about living organisms and biology. On the contrary, the discovery of the double helix and everything that has happened in molecular biology since lie squarely within the dual paradigms of the organism as a machine and as the kind of machine that will be understood better and better as we explain the workings of its smaller and smaller parts (in other words, of reductionism).

Before I get into the scientific part of the story, I want to try to answer another sociological question: What can we learn from Watson's *Double Helix* about the way science is done and, particularly, about how women and men relate to scientific work? This is not a trivial question because the book was a best seller and is still on reading lists in many high schools. So, let us see how Watson introduces his four principal characters: himself, Francis Crick, Maurice Wilkins, and Rosalind Franklin. (Watson also acknowledges Linus Pauling as a main player, but Pauling comes into the book quite late and is never rounded out like the others, who enter in the first half dozen pages.)

We meet Watson in the Introduction, climbing in the Alps, where he is hailed by a London colleague as "Honest Jim," the phrase he originally intended for the title of the book. (As I recall, he didn't use it because of its similarity to Kingsley Amis's *Lucky Jim*.) Crick is introduced in the first sentence ("I have never seen Francis Crick in a modest mood") and, by the fourth, is likened to two giants of physics, Rutherford and Bohr. We have to wait for Wilkins till the second chapter, only three pages further on, where we learn that "at this time [fall 1951] molecular work on DNA in England was, for all practical purposes, the personal property of Maurice Wilkins, a bachelor who worked in London at King's College. Like Francis [Crick], Maurice had been a physicist and also used X-ray diffraction as his principal tool of research."

So the stage is set with three male principals: Watson, a bit frivolous and problematical, although clearly extraordinary; Crick, loud, ebullient, and brilliant; Wilkins, serious and a bit musty. Wilkins's qualities are emphasized within another paragraph by the statement that "Maurice continually frustrated Francis by never seeming enthusiastic enough about DNA." At the end of this paragraph we meet Franklin: "Moreover it was increasingly difficult to take Maurice's mind off his assistant, Rosalind Franklin."

Here I must pause at once to explain that Franklin was not Wilkins's assistant. The two had independent and equivalent appointments in the laboratory of John (later, Sir John) Randall, professor at King's College. Furthermore, although Franklin was new to this laboratory and Wilkins had been there for some time, the reason Randall invited her was because she had more experience with x-ray diffraction than Wilkins. Randall's correspondence with Franklin had led her to expect that she would be working with DNA on her own, not with Wilkins.

But let us go on with Watson:

> Not that he was at all in love with Rosy, as we called her from a distance. Just the opposite. . . . Maurice, a beginner in X-ray diffraction work, *wanted some professional help* and hoped that Rosy, a trained crystallographer, could speed up *his* research. Rosy, however, did not see the situation this way. She claimed that she had been given DNA for her own problem and would not think of herself as Maurice's assistant [my italics].

Then follows a disparaging account of Franklin's physical appearance—her hair, clothes, and her lack of femininity and grooming. The description ends with this statement: "Clearly Rosy had to go or *be put in her place.* The former was obviously preferable because, given her belligerent moods, it would be very difficult for Maurice to *maintain a dominant position that would allow him to think unhindered* about DNA" (my italics).

These passages speak for themselves and require only one amplification: No one—family or friends—ever addressed Franklin as Rosy. The name is part of the stageset in which this dowdy (she happened not to be that), petulant, uppity bluestocking claimed, as her own, work that rightfully belonged to Wilkins.

As the story goes on, this picture is embroidered until it

becomes clear that real science—good science—is done by bright, ambitious men who relate to women in one of two ways. Beautiful, charming women offer a delightful escape from important and serious concerns, such as finding the key to life. But women can also be a damned nuisance if they aren't sufficiently helpful and especially if they try to be scientists and have their own ideas. In that case, they can stem progress and stop science dead.

I want to pause for a moment and compare Watson's description of how science gets done with the descriptions in two well-known, earlier stories: Sinclair Lewis's *Arrowsmith* and C. P. Snow's *The Search*. Written in the 1920s and 1930s, these books are very different from *The Double Helix*. They paint scientists as idealists, dedicated to the search for Knowledge and Truth, and much less interested in worldly success and honors than Honest Jim. In fact, when the hero of Snow's novel realizes that he has begun to do his scientific work mostly in order to attain status and success, he decides to stop being a scientist. And Arrowsmith ends up doing his research in a small private laboratory in the woods, away from worldly distractions.

All three books agree on one thing: the roles of women. In both *Arrowsmith* and *The Search* women exist to nurture and sustain the hard-working scientist and to offer him much-needed relief from his important work. In those capacities women play major roles. They are significant, strong, and sorely missed by the scientist-heroes when they leave or die. The three times Lewis mentions professional women it is to scoff: once when he describes Arrowsmith's female fellow medical students as "virginal and unhappy coeds"; another time, when they are "emotional and frightened"; and a third, when they "shudder" while watching the great professor—Arrowsmith's role model— inject a virulent strain of bacteria into a guinea pig.

The Search is even more interesting than *Arrowsmith* because it contains several strong and important women but only a single hint that women, too, are scientists. This hint appears when a male scientist, who makes a hobby of seducing professors' wives, promises a friend who tries to help him further his career not to seduce "a scientist's wife. Not one. Not even a woman scientist." That is the only mention of women scientists in this novel about British x-ray crystallographers, a professional group that is unusual among scientists because it includes many successful and distinguished women.

So we are in the customary bind: When women actively participate in a field that is not stereotypically female—and do so not in the accepted roles as cleaning women, technicians, secretaries, or wives—they are either maligned or ignored and written out of the story.

Scientific

I shall now evaluate Franklin's contribution to the elucidation of the structure of the double helix, using as sources her own and other people's publications. The main bibliographic references are available in Robert Olby's (1974) *The Path to the Double Helix*, in Sayre's book, and in my review of it (Hubbard, 1976).

In 1951, Wilkins was given a sample of DNA, and he and a graduate student, R. G. Gosling, took a first, quite good x-ray diffraction picture of it. This is the picture he showed Watson when they met in Naples. Watson got excited, and because he was not enthusiastic about his work in Copenhagen, he decided to move to Cambridge, England, to learn about x-ray crystallography and work on the structure of DNA. There he met Crick in the fall of 1951. Crick, at thirty five, was working for his Ph.D. at the Cavendish Laboratory; Watson, twenty three, already had his doctorate. Both have recalled that they were immediately struck by how similarly they thought about biology. Both considered the structure of DNA to be the most important problem in biology and wanted to work on it. Unfortunately Crick, who was familiar with x-ray crystallography, was in the middle of quite different work which he needed to finish for his Ph.D. thesis. So Watson had to learn quickly.

As we have seen, at this time Wilkins was working on DNA at King's in London and his chief, Randall, had invited Franklin, an expert in x-ray analysis, to come and build up the x-ray diffraction unit there. Franklin had spent the preceding several years in Paris doing x-ray analysis of different forms of carbon (coal) and wanted to turn her attention to biological substances.

She therefore was glad to accept a research fellowship in Randall's biophysics unit, which she joined in January 1951. Wilkins was out of the country when Franklin arrived, and her work was under way by the time he returned. It is not clear what each thought their relationship was supposed to be, but Sayre writes that they took an instant dislike to each other.

Perhaps this is the time to ask whether Franklin or Wilkins was just plain difficult to get along with. Watson would have us believe Franklin was. My own inquiries suggest that she was argumentative and at times defensive and prickly. But she collaborated successfully with her colleagues in Paris, with Gosling at King's, and, after leaving King's, with Aron Klug and other scientists at Birkbeck College in London. Similarly Wilkins has had many collaborators, although usually not women. Sayre is probably right when she says that this was an unfortunate collision of two very different, and apparently incompatible, temperaments. No doubt the situation was also exacerbated by antiquated sexist practices at King's that excluded Franklin from relaxed socializing with her male colleagues over morning coffee or afternoon tea, which were served in separate and unequal combination rooms. (Why, we ask with Virginia Woolf, are women always served boiled beef and prunes, while the men feast on partridges and wine?) Whatever the reasons, tensions between Franklin and Wilkins developed almost immediately.

Franklin began by building a high resolution x-ray camera to study the structure of DNA with Gosling. Wilkins worked on DNA with A. R. Stokes, and there was little communication between the two pairs, even though they were housed in the same unit at King's. Within a few months, Franklin discovered that, depending on its water content, DNA forms two kinds of fibers with quite different x-ray diffraction patterns. She called the dry form A, the wet form B, and decided to begin her structural analysis with the A pattern, which showed more detail, planning to go on to the B pattern later. In retrospect, this was a misjudgment, but one that made sense to her at the time.

In November 1951, when Watson had been in Cambridge about six weeks and as yet understood little about x-ray diffraction, he went to King's to hear Franklin give a seminar about her work on DNA. We learn from *The Double Helix* that he did not take notes, ruminated on Franklin's lack of sex appeal, misunderstood and misremembered much of what she said, and communicated his recollections to Crick the next day. Crick got intrigued and started to make calculations. On this basis the two of them quickly built their first DNA model and invited the group from King's to come see it. Franklin instantly pointed out that the model could not be reconciled with data she had presented at her seminar and would have nothing further to do with such ignorant and idle speculation.

This encounter had important consequences. For one thing, it made Sir Lawrence Bragg, who headed the Cavendish Laboratory in Cambridge, order Watson and Crick to stop working on DNA because the group at King's was in the midst of doing it. There were also subtler consequences for Watson and Franklin. Watson had been shown up by a woman and one whom he thought of as Wilkins's technician. Franklin seems to have decided that Watson was a clown and stopped taking him seriously. After all, he had sat through her talk and promptly gone off and built a highly speculative model that ran counter to her data and was obviously wrong—not her idea of how to do science. She saw no reason to shift over to model building and proceeded with her systematic analysis of the A pattern.

Watson and Crick, however, were obsessed by the realization that Linus Pauling and his colleagues at Caltech had recently "cracked" the structure of proteins by careful and imaginative construction of three-dimensional models. Yet there is an important difference between the successful bout of model building by Pauling's group and what Watson and Crick had tried to do. Prior to building their models, the group around Pauling had systematically studied the structures of all the component molecules and established a large body of information regarding their bond lengths, bond angles, and possible configurations. Watson and Crick did not even bother to find out what was known about the molecules with which they were trying to build their models. Erwin Chargaff, one of the pioneers in nucleic-acid chemistry, when visiting Watson and Crick in the summer of 1952, found they did not even know the chemical formulas of the so-called purine and pyrimidine bases that compose DNA, much less their spatial configurations. Some six months later Jerry Donohue, another chemist, had to instruct them that these bases can exist in either of two configurations, one of which is considerably more stable than the other and turned out to be the one the bases assume in DNA.

Late in 1952, Watson and Crick heard that Pauling thought he had figured out the structure of DNA and was writing a paper about it. Through his son, Peter, they got a copy of Pauling's manuscript and noticed that he had made a rather obvious mistake. With that they decided that all bets were off. If Pauling was going after DNA, then it no longer belonged to King's; so why shouldn't they work on it?

About this time, Wilkins showed Watson Franklin's best x-ray diffraction picture of the wet (B) form of DNA, which clearly showed it to be helical. Meanwhile their Cambridge colleague, Max Perutz, gave Crick and Watson a copy of a privileged research report from the King's group to the Medical Research Council in which Franklin summarized her most recent findings including the spacings of the critical reflections on the x-ray diffraction pattern of the DNA fibers. Any correct model of DNA had to account for these spacings and for their relative intensities. So, seeing themselves in a race against Pauling and with Franklin's data against which to check plausible models, Watson and Crick went into a frenzy of model building. Within a few weeks they arrived at the now-famous structure, wrote it up, and sent their manuscript to Wilkins.

This was the first Wilkins knew that Watson and Crick had gone back to working on DNA. He and Franklin immediately accepted the new model and on April 25, 1953, there appeared in *Nature* the historic triad of papers. The first, by Watson and Crick (1953), described the structure; the second and third, by Wilkins, Stokes, and H. R. Wilson (1953) and by Franklin and Gosling (1953), offered supporting x-ray evidence. The papers made a big splash, and Watson and Crick were instantly famous.

At the time, Franklin was in the process of moving from King's, which she hated, to J. D. Bernal's laboratory at Birkbeck College, where she did distinguished work on the structure of tobacco mosaic virus virtually until the day she died of cancer in 1958, at the age of thirty seven. She never knew that the crucial features of her "supporting" evidence were in Watson's and Crick's hands before they began to construct their model.

Watson, Crick, and Wilkins shared the Nobel Prize in 1962. And the matter probably would have ended there had Watson not published *The Double Helix* in 1968. Until then no one, except Watson and Crick, knew that they had used Franklin's data—unbeknownst to her—to guide their formulation of the DNA structure.

Before continuing the story, it will help to summarize the main features of the Watson/Crick model of DNA. DNA is pictured as composed of two ribbons, each of which consists of a long, invariant, alternating sequence of molecules of a sugar (deoxyribose, hence the D in DNA) and phosphate: -sugar-phosphate-sugar-phosphate-. The ribbons have a polarity that makes it possible to distinguish one end from the other, and two of them, running in opposite directions (hence, head to tail), are wound into a double helix. The two ribbons are connected at regular intervals by horizontal rungs, each of which is formed by a pair of flat discs, the purine and pyrimidine bases. There are four different bases in DNA: adenine (A), guanine (G), thymine (T), and cytosine (C). The two purines, A and G, are the larger units; the two pyrimidines, T and C, smaller. To fit inside the helix, A must always pair with T, G with C. This feature of the helix explains observations Chargaff and his colleagues at Columbia University made several years earlier, when they showed that although DNAs from different organisms contain different proportions of the bases, the amount of A is always the same as that of T, and the amount of G the same as that of C.

Watson describes the DNA structure as "a spiral staircase with the base pairs forming the steps." The helix can be very long, so that a DNA molecule can contain many thousands of base pairs, arranged in an almost infinite number of sequences. Therefore there are a huge number of different DNAs. This diversity is necessary if the millions of different genes that mediate the many inherited characteristics of organisms are DNAs with different base sequences. That this diversity can be encompassed in such a simple structure is one of the reasons the model won immediate, wide acceptance.

Watson and Crick's paper in *Nature* specified all the critical dimensions, such as the pitch and diameter of the helix and the number of turns between repeats. Nowhere do they acknowl-

edge that they obtained them all from Franklin's data and calculations. Furthermore, Franklin was the one to insist as early
as 1951 that the sugar-phosphate backbone had to be on the
outside of the fiber, with the bases pointing in, because that was
the most reasonable explanation of why DNA fibers readily
take up water. She had also suggested that the bases were probably held together by hydrogen bonds. Both features are part of
the Watson-Crick model, again without acknowledgment.

In a longer paper that Crick and Watson (1954) submitted to
the *Proceedings of the Royal Society* of London in the latter part of
the summer, they wrote: "We have only considered such structures as would fit the preliminary X-ray data of Wilkins, Franklin
and their co-workers. Our search has so far yielded only one
suitable structure." So here they do say it, but this fuller account
is rarely referred to or read. And it, too, does not acknowledge
that they had seen Franklin's privileged report without her
knowledge and that it contained her calculations of all the critical parameters.

Crick and Watson were the ones who saw how to put other
people's data and suggestions together and come up with the
DNA structure, a first-rate, imaginative achievement. They
would have been no worse off had they acknowledged their
debt to Franklin. It is therefore interesting to look at what they
did acknowledge. In the *Nature* paper, they say: "We are much
indebted to Dr. Jerry Donohue for constant advice and criticism, especially on interatomic distances. We have also been
stimulated by a knowledge of the *general nature* of the unpublished experimental results and ideas of Dr. M. F. H. Wilkins,
Dr. R. E. Franklin and their co-workers at King's College, London" (my italics).

The *Royal Society* paper ends: "We are most indebted to Dr.
M. F. H. Wilkins both for informing us of unpublished experimental observations and for the benefit of numerous discussions. We are also grateful to Dr. J. Donohue for constant
advice . . . and to Prof. A. R. Todd, FRS, for advice on chemical matters and for allowing us access to unpublished work."
There follows a remarkable final paragraph in which "one of us
(J. D. W.)" acknowledges "hospitality," "encouragement," and
general indebtedness to Sir Lawrence Bragg, J. C. Kendrew,

M. F. Perutz, and S. E. Luria—a list of important and powerful scientists, all of them then or future Nobel laureates. Not a word about Franklin.

Implications for Science Making

I want to emphasize that I do not think the way Watson presents Franklin is idiosyncratic—perhaps a bit exaggerated, but not that unusual. Seeing a woman work in a laboratory, Watson is not alone in assuming that she must be somebody's technician. As late as 1970, some two years after the publication of *The Double Helix*, Pauling, who had good reasons of his own to be leery of the scurrilous innuendo in the book, referred to Franklin's pictures of DNA as "Wilkins' X-ray photographs."

A question that is often asked is how close Franklin was to solving the DNA structure. Obviously there is no answer. Franklin was working systematically and making good progress. She was intent on exhausting the possibilities of traditional x-ray analysis before getting involved with model building, which she considered needlessly speculative. Had she continued to work on DNA, she undoubtedly would have solved the structure. But because she hated her situation at King's, Franklin had decided to move to Bernal's unit at Birkbeck College. Sayre tells us that Randall made her agree, once she had left, "to stop thinking about DNA entirely" (p. 168), to which Franklin responded quite appropriately, "But how could I stop thinking?" (p. 214). So it is probably historically correct that she would not have ended up solving the structure—if such speculation can be called history. Whether and how much the fact that Watson was breathing down her and Wilkins's necks exacerbated the strains between them, no one will ever know.

At the point at which Watson and Crick began their climactic round of model building, a number of important facts were known about DNA: (1) Wilkins and Stokes and also Franklin and Gosling had shown that at least the B form was helical; (2) Franklin's conclusion that the sugar-phosphate backbone lies on the outside meant that the bases must fit into a regular, repeating pattern *inside* the helix; (3) her x-ray data supplied the critical parameters against which to test possible models; (4) Chargaff's rules regarding the equivalence of the amounts of A and T and of G and C contained in DNA set further limits. With all these constraints as guideposts, the structure was bound to be solved before long.

The story of the search for the structure of DNA has implications beyond those for women in science. It is sometimes said that Watson and Crick's race for the double helix and its celebration in Watson's best-selling book have lowered the moral tone of science and escalated its frenetic competitiveness, particularly in molecular biology. I am skeptical. After all, science reached its *man*hood during the heyday of industrial capitalism, when competition was hailed as the road to success in a system that was claimed to be meritocratic. Like others growing up within the liberal ideology, scientists operate with the assumption that competition sorts the chaff from the wheat and that ability, not accidents of birth, enterpreneurial skills, or ruthlessness, determine success. Since the discovery of the double helix, the number of young people entering science has increased greatly but the number of available positions has not. I attribute the deterioration in the social relationships and mores—the enormous competitiveness and secretiveness that poison the contemporary scene for many aspiring young (and not so young) scientists—on this larger context more than on the winner-takes-all morality exemplified in Watson's race for the Nobel. This is not to deny that the overwhelming success of that race and the power and prestige it quickly brought the until-then-obscure Watson and Crick probably helped give it respectability.

In this connection, it should not go unnoticed that Bragg, who less than two years earlier had summarily ordered Watson and Crick away from DNA, was far from disapproving when they "won" DNA for Cambridge (which was still smarting from the "loss" of protein structure to Pauling). Bragg went further and wrote the foreword to *The Double Helix*—a decided coup for Watson in the face of vociferous disapproval of the manuscript by the chief characters in it, including Crick, who threatened a lawsuit. In his foreword, Bragg expresses "deep satisfaction" at the "due recognition . . . given to the long, patient investigation by Wilkins at King's College (London) as well as to the brilliant and rapid final solution by Crick and Watson at Cambridge" by the Nobel committee's decision to split the prize between them. Once again not a word about Franklin.

A final point: As epigraphs to the concluding chapter in *The Path to the Double Helix,* Olby (1974) has two quotations, one from Crick, the other from Michael Polanyi, a philosopher of science who began his career as a physical chemist. The quotation from Crick begins: "The ultimate aim of the modern movement in biology is in fact to explain *all* biology in terms of physics and chemistry." The rest expands on this reductionist paradigm and ends: "It is the realization that our knowledge on the atomic level is secure which has led to the great influx of physicists and chemists into biology." Placed right up against it, the quote from Polanyi ends with the words: "Now the analysis of the hierarchy of living things shows that to reduce this hierarchy to ultimate particulars is to wipe out our very sight of it. Such analysis proves this ideal to be false and destructive."

This is the parting of the roads that we confront in modern biology. The reductionists call their opponents pessimistic; the holists call the reductionists blind, if not destructive and dangerous. Nothing illustrates the dichotomy more clearly than the controversies that have arisen over genetic engineering of bacteria, plants, and animals, and more recently over the Human Genome Initiative, the projected effort to identify the genes and establish the sequence of all the bases on all twenty-three human chromosomes (see chapter 6). All these are practical implementations of the reductionist belief that organisms are merely the living manifestations of their DNA. This is the side on which Watson, Crick, and many other molecular biologists line up. On the other side are people who see organisms as more than the sum of their parts, who believe that connections and process cannot be understood by isolating things or events, who distrust reductionist oversimplifications.

We shall see more of this debate in the next two parts of the book. For now, I want to emphasize once again that in the real world everything is connected with everything else, that "isolated variables" are figments of the imagination. "To light a candle is to cast a shadow," writes Ursula LeGuin (1968) in *A Wizard of Earthsea*. This does not mean that we should abandon all attempts at scientific analysis. But we must keep our perspective about what such analyses can tell us. We cannot expect the science of atoms and molecules to teach us what life is.

PART TWO

What Do We Know?

We saw in Part I that by erasing the observer and pretending to objectivity, scientists obscure the subjectivity of their reading of nature. And by subjectivity I do not mean personal idiosyncrasies of individual scientists but the unacknowledged class, gender, and racial affiliations and world-view scientists represent and express in their work. The chapters in Part II provide examples of the way implicit ideological commitments inform the scientific understanding of genetics and evolutionary processes and of concepts of human nature, women's biology, sex differences, and sexuality.

This way of framing the issue of how the dialectic between objectivity and subjectivity affects our understanding of nature puts this book into the middle of an ongoing feminist debate about the possibility of generating a "better" science, not just a different one (Harding, 1986). Try as I will to adopt a relativism that acknowledges different ways of interpreting nature but regards all of them as stories of equal weight, the scientist in me insists on trying to distinguish between more and less accurate representations of nature and social reality.

I believe that the effort to specify the societal affiliations of the scientific observer and to identify her or his viewpoint will result in increasingly accurate interpretations of nature and society and therefore in improved scientific theories and descriptions. In addition, I would hope the new insights will be applied conscientiously and cautiously because practitioners will have to take explicit account of the context in which they make their observations. They will have difficulty pretending to objective detachment when their scientific findings are implemented in destructive ways.

At the moment, I can think of few examples of this kind of contextualized scientific work. As I mentioned before, one was the way the Boston Women's Healthbook Collective (1984) generated *The* New *Our Bodies, Ourselves,* a process in which the research and writing involved many women, selected to encompass diverse experiences and points of view. These women conducted interviews, researched the relevant literature, and discussed their findings together. Another is the research that went into *Drylongso,* a book in which the black anthropologist John Langston Gwaltney (1980) weaves together interviews he conducted with black women and men about their experience of living in the United States. A third is the research Emily Martin (1987) describes in her book *The Woman in the Body.* She and her co-workers gathered information from 165 women of different classes, ages, ethnic groups, and races living in Baltimore in order to record and analyze the ways they experience puberty and menstruation, pregnancy and childbirth, and menopause and aging.

The authors of these books carefully describe their research methodology and their relationship to people they interviewed. Their exploration of their subjectivity produces a self-conscious objectivity—an ongoing juxtaposition of what they hear and who they are—that allows for a more accurate assessment than usual, as well as a direct and lively account, of their observations. It acknowledges the boundaries beyond which the results should not be generalized and produces a science that non-experts can enjoy reading and can even imagine doing.

This kind of research opens up the possibility of a science practiced not just for the people but by the people (Benston, 1986). For example, Gwaltney initiated what he calls "folk seminars" in which he listened to groups of black people as they reflected together on their lives in white America as well as on his research project and made suggestions about how social science could, and should, help them. Martin, in turn, is aware of the changes in outlook that were caused by the introspection that the study elicited among the research "subjects" as well as among the researchers and of the implications this process has for the practice of social science.

I offer these examples as concrete alternatives to the science I describe (and criticize) in the next six chapters because without other models traditional, "objective" science may seem the only science possible.

Some parts of these chapters are more technical than the rest of this book. I urge readers without a scientific background not to be put off. I have tried to make the scientific discussions brief and clear. But if you do not want to get into them, don't just stop reading the chapter. Skim ahead. You can get the gist of the argument anyway.

6

Genes as Causes

GENETICS is the systematic description of hereditary mechanisms, but to a large extent it is also a reading into nature of the ideologies of hereditarianism and individualism that were dominant during the period when it was invented. Hereditarianism dominated the social thinking of Spencer and the other Social Darwinists and was evident in the novels of Dickens, George Eliot, and Thackeray, as well as in the scientific thinking of Francis Galton and the eugenicists, about whom I will have more to say in later chapters.

This ideology has been the impulse behind a great deal of genetic experimentation in the expectation that genetics can be useful for purposes of social and genetic engineering. For example, H. J. Muller ([1913] 1962) wrote near the start of his scientific life: "The intrinsic interest of these questions [about heredity] is matched by their extrinsic importance, for their solution would help us to predict the characteristics of offspring yet unborn and would ultimately enable us to modify the nature of future generations."

This goal is echoed by many present-day advocates of genetic engineering. For example, the American theologian Joseph Fletcher (1974, p. 4) writes: "The intention of genetic engineering is to locate and alter the genes which cause defects in superior people. . . . Discoveries like genetic coding make the breakthroughs of old-fashioned hard technologies seem like child's play."

Nineteenth-century biologists were interested in exploring the intrapersonal "causes" of what Galton termed "hereditary genius" and of numerous less desirable attributes they considered equally innate, such as "pauperism." They hypothesized that these characteristics were mediated by a hereditary substance inside cells, which different ones of them denoted by different names. Galton called it "stirp," Karl Nägeli "idioplasm," and August Weismann "germ plasm." Some of them assumed that the genetic substance that is transmitted to successive generations is made up of particles, which Darwin called "gemmules," Weismann "ids," Hugo De Vries "pangenes," and finally W. Johannsen "genes"—the name that stuck. The concepts or things represented by these various terms were not the same. But the fact that all of them were invented to denote hypothetical hereditary particles suggests that there was a strong ideological need to assume the existence of material substances, often particles, located within individuals, that transmit traits from one generation to the next. And their invention predates what can legitimately be called a science of heredity—that is, genetics.

Gregor Mendel, who is quite properly credited as the father of genetics, had a more limited goal than that of these biologists. In his classic paper in 1865, Mendel did not go beyond formally designating "characters" (observable traits). He suggests only once that these might be correlated with hypothetical "factors" inside his pea plants. Other scientists assumed that observable traits and the ways they are passed on to, or changed between, successive generations must be associated with material substances within the organism. For example, the German embryologist August Weismann wrote in 1893 (p. 11): "The phenomena of heredity among higher organisms are connected with a definite substance . . . localized . . . in the nuclear substance of the germ cells."

Although chromosomes were observed shortly after the turn of the century, genes continued much longer as purely theoretical constructs. But the materialist and reductionist impulse that led physicists to describe matter in terms of atoms and, later, subatomic particles led biologists to assume quite early that inheritance is mediated by intracellular, hereditary particles. As the American geneticist Thomas Hunt Morgan explained in 1926 (p. 1): "In the same sense in which the chemist posits invisible atoms and the physicist electrons, the student of heredity appeals to invisible elements called genes."

After the rediscovery of Mendel's laws around 1900, his

"factors" became the genes. During the following decades the hypothetical genes became concrete objects—pieces of chromosomes—and finally, in 1953, molecules of double-stranded DNA. But it would be well to keep in mind that the geneticist L. J. Stadler admonished as late as 1954 that "our concept of the gene is entirely dependent on the occurrence of gene mutations," that is, on changes in inherited traits. This is an important point that is worth exploring further.

What made it possible for Mendel and the later geneticists to establish formal laws of heredity and explore the mechanism of inheritance experimentally is the fact that traits can change from one generation to the next, not that they stay the same. Changes are what made it possible to sort and name specific traits out of the overall similarities that get passed on from one generation to the next. The British geneticist J. B. S. Haldane (1942) put it this way (p. 11): "Genetics is the branch of biology which is concerned with innate differences between similar organisms. . . . Like so many other branches of science, genetics has achieved its successes by limiting its scope. Given a black and a white rabbit, the geneticist asks how and why they differ, not how and why they resemble one another."

Johannsen, the man who coined the word *gene*, also invented the terms *phenotype* and *genotype* to distinguish an organism's outward appearance, which he called its phenotype, from hypothetical, correlated internal factors, its genotype. These two terms have often been used as though the genotype determines the phenotype, even though many scientists have stressed that phenotypes are generated by an ongoing, complex, and mutual interplay between genotypes and environments. When by genotype one means genes, then the environment includes not only what happens outside the organism but also the many metabolic reactions that go on in cells and between them. Therefore I want to stress from the start that nothing in the concept of factor, gene, or genotype necessarily implies a causal line from it to characters or traits, as the previous quote from Fletcher would suggest.

Genes (DNA) differ from one another in their base sequence, as I explained in chapter 5. Because the differences among them are quite specific, genes introduce specificity when they interact with other molecules. But specificity is also imparted by other molecules, among them RNA (which is another type of nucleic acid involved in translating the molecular properties of genes into traits), proteins, and even some carbohydrates and lipids (which are a class of fats). Many of the processes that take place in organisms and during the interactions in which organisms engage their environments can mediate specificity.

To repeat, hereditarianism, the belief in the biological inheritance of socially important traits, was part of all the major trends in nineteenth-century thought. It was fundamental to the thinking of the conservative followers of Thomas Malthus, to liberal beliefs in meritocracy, and to evolutionist hopes for human betterment. The scientific question was how traits ("characters") were passed from one generation to the next. That the answer was found in hypothetical particles was not surprising at a time when scientists demonstrated that matter and energy consist of particles and that particles carry diseases. In fact, similar fallacies are involved in trying to explain how bacteria "cause" disease and how genes "cause" traits. In both cases, the power of the particles is exaggerated, while the contributions of the systems in which they operate are undervalued or ignored. This belief in the superior explanatory power of the smallest elements in a system is reductionism, which we have discussed previously.

What Mendel's laws can and cannot tell us has been explained in a book published collectively by a group of French biologists who call themselves Agata Mendel (1980). (They invented Agata by analogy to Virginia Woolf's story in *A Room of One's Own* about the life of Shakespeare's imaginary sister, Judith.) This book raises important issues that are usually overlooked and I want to go over them in some detail.

Mendel's Laws

As I said before, Mendel's laws, which are basic to modern genetics and molecular biology, are about change and difference, not about likeness. But so urgent is our need to explain why "like father, like son" that experiment after experiment is interpreted as though we could.

Mendel devised his experiments to explain certain regularities observed when dissimilar individuals are crossed to produce hybrids. He chose for his experimental organism the sweet pea (Pisum) because it fulfilled certain basic conditions: It exhibits characters which "stand out clearly and definitely" and do not differ in a "more or less" way (Mendel, [1865] 1950, p. 4). In other words, he specifically selected differences in kind rather than of degree—qualitative, not quantitative, differences. He eventually selected seven pairs of characters, an example being round, smooth seeds as against angular, wrinkled ones.

A simple convention governs the way successive generations are denoted in Mendelian genetics. P_1 signifies the parent generation, which produces the first generation of offspring, called F_1. Successive generations of offspring, produced by breeding members of the preceding generation together, are called F_2, F_3, etc. In other words, the P_1 generation produces the F_1 generation, F_1 produces F_2, and so on.

Mendel first made sure that both parental lines (P_1) bred true, and that crossing the P_1 plants produced the same kind of hybrid (F_1), no matter which parent strain provided pollen and which eggs. He also took care to select pairs of characters whose hybrid (F_1) closely resembled one of the P_1 strains rather than looking like a mixture of the two parental lines. He called "those characters which are transmitted entire, or almost unchanged in the hybridization . . . dominant, and those which become latent in the process recessive" (p. 8). He explained that he used the term *recessive* because although such characters "withdraw or entirely disappear in the hybrids, [they] nevertheless reappear unchanged in their progeny" (p. 8).

Mendel established his laws by examining four successive generations: P_1, F_1, F_2, and F_3. He showed that when the F_1 hybrids are bred with each other, in the F_2 generation "there reappear together with the dominant characters, also the recessive ones . . . in the definitely expressed proportion of three to one" (p. 9). He then showed that those F_2 plants that exhibited the recessive trait bred true, like the recessive P_1 plants, and continued to breed true for however many generations he bred them. But the F_2 plants that exhibited the dominant character divided into two groups: two-thirds behaved like the F_1 hybrid and one-third bred true, like the dominant plants in the P_1 generation. He therefore concluded that the appearance of a $3:1$ ratio in the F_2 generation resulted from what was in fact a $1:2:1$ ratio, in which one part behaved like the dominant P_1 plants, two parts like the hybrid F_1 (which looked like the dominant P_1 plants but didn't breed true), and one part like the recessive P_1 plants.

Although he denoted the dominant pure line as *A*, the recessive as *a*, and the hybrid as *Aa*, he used these symbols as formal representations that were not meant to signify something inside the plant that caused or produced the character. He departed from the descriptive use of the term *character* at only one place, when he assumed the presence of identical "factors" in the pure lines and the hybrids in order to explain that hybrids (*Aa*) can give rise both to pure *A* and pure *a* lines.

(R. A. Fisher has suggested that Mendel, who was trained in mathematics and physics, probably had his theory of hybridization worked out before he began the extensive, painstaking series of breeding experiments (Olby, 1966). It is plausible that Mendel planned his experiments as decisive tests of a worked-out hypothesis rather than that his vast accumulation of data, collected over seven years, "led" him to his laws.)

Following the rediscovery of Mendel's paper just after the turn of the century, the new geneticists soon transmuted his hypothetical factors into genetic particles, or genes, that were located on chromosomes and said to cause traits by their chemical action. The nature of this action was not specified until mid-century, after the chromosomes, hence genes, had been shown to contain the bulk of a cell's DNA. As we saw in chapter 5, the structure of DNA was elucidated in 1953, which made it possible to explain how DNA is copied when cells divide and to explore some of its other chemical activities. During this half century, the laws by which Mendel formally described how hybrids can give rise to differences were imperceptibly and without conscious intent interpreted as though they described how like characters are passed from parents to offspring.

The fact that the pattern of inheritance of an observable, qualitative difference between two organisms of the same species can be described by Mendel's laws has led scientists since Mendel to assume that the difference is mediated by different forms of the same gene, called *alleles*. But this assumption is not equivalent to saying that the gene, in any of its allelic forms, generates or causes the character in question. Nothing we have learned since Mendel's time about how genes function justifies the conclusion that genes cause traits or control development. They make significant contributions to both, but so do lots of other substances, including other genes.

Let us take hemoglobin, the protein that gives blood its red color and carries oxygen around the body. Normal and sickle-cell hemoglobin are a pair of "characters" whose inheritance follows Mendel's laws, with normal hemoglobin dominant over sickle-cell hemoglobin. These two kinds of hemoglobin molecules are known to differ by a single amino acid. (Amino acids are the small molecules of which the much larger proteins are built.) The difference can be mediated by a change, a mutation, in a single base pair in the DNA sequence, or gene, involved in hemoglobin synthesis. It is correct to say that, all other things being equal (which they never are outside of controlled laboratory experiments), the difference in the two forms of DNA exerts a decisive influence during the synthesis of the two forms of hemoglobin. It is not correct to assume that either gene causes one or the other form of hemoglobin to be synthesized.

Hemoglobin synthesis, like the synthesis of any other complex molecule, involves a battery of reactants and energy sources that must come together at the appropriate times and under appropriate conditions, among them appropriate genes (DNA). If the DNA in the "hemoglobin gene" is changed, a different hemoglobin will be formed. Or perhaps hemoglobin will not be formed at all if the change in the DNA is too great or of the wrong sort. But if one of the other genes that must get into the process is changed, or a critical concentration of one of the essential small molecules, or one of the enzymes, or the temperature, or the pH, or . . . , normal hemoglobin also may not be formed. Indeed the cell, or even the organism, may not survive at all.

Take another familiar Mendelian character—the Rh blood antigen, which is present in Rh positive people (Rh+), but lacking in Rh negative ones (Rh−). The Rh+ trait is dominant over Rh− and, by inference from Mendel's laws, one can assume that the difference involves two forms of one gene. This assumption in no way allows one to conclude that the "Rh+ gene" causes Rh antigen to be formed or that it directs its synthesis.

The point to stress is this: In a complex system of reactions, such as protein synthesis, which requires many components and conditions that must work together in nonadditive ways and that often are interdependent, it is wrong to single out any one substance or event as the cause.

DNA as Information

Some biologists insist on the primacy of DNA, while granting that the biosynthesis of a protein, such as hemoglobin or the Rh antigen, is complex and requires interactions among many molecules, subcellular structures, and processes. They argue that it is appropriate to assign a commanding role to DNA because it contains the information that imparts specificity. But as Barry Commoner (1968) pointed out, the enzymes involved in the synthesis of RNA and DNA, which, like all enzymes, are proteins, also impart specificity. He concluded "that the biological specificity of inheritance originates in no one molecule but in a complex circular network of molecular interactions in which various DNA, RNA and protein agents participate." Questioning the workability of genetic engineering, he went on to say: "If . . . the specificity transmitted in inheritance is determined by a multimolecular system vastly more complex than a DNA molecule, any promise to control inheritance by chemical manipulations of DNA is likely to be illusory."

Most molecular biologists, if pressed, grant that there are multiple sources of specificity and control in biosynthetic systems and that the critical control points undoubtedly change over time. Despite this admission, they give primacy to DNA because they insist that information usually passes in only one direction, from DNA to RNA to protein. But the whole point of a complex metabolic system is that at one time or another all the parts must talk to one another.

Donna Haraway (1979) has pointed out that at the end of World War II integrated ways of thinking about cellular interactions and relationships inside organisms, such as those

Walter B. Cannon (1939) expressed in his book *The Wisdom of the Body,* were replaced by analogies from information theory. This was when "feedback circuits" and "information flow" entered biology. The reasons for this change undoubtedly included the wartime flow of biologists into operations research and the postwar flow of disillusioned physicists into biology. These kinds of changes in outlook brought into genetics ways of thinking about information, coding, and control that are considerably more mechanistic, reductionist, and hierarchical than biochemical thinking was before the war.

In the late 1950s, molecular biologists formulated what Crick called the Central Dogma—that genes act by virtue of the fact that DNA defines the structure of RNA, which in turn defines the composition and structure of proteins: in shorthand, DNA→RNA→proteins. At that time, scientists assumed that the base sequence of a gene (DNA) completely specifies the sequence of amino acids of the protein in whose synthesis it participates (or, as they would say, "for which it codes"). In addition, they assumed that the amino-acid sequence of a protein determines its three-dimensional structure and, by implication, its function. Hence, it seemed logical that the shape of hemoglobin, its color, and the way it transports oxygen in the blood all are determined by "the hemoglobin gene."

We now know that this "dogma" is not true. In organisms other than bacteria, the base sequence of a gene (DNA) is not translated literally into the amino-acid sequence of the protein in whose synthesis it participates. Some portions of the DNA "template" are not used at all, while others are placed into new juxtapositions. These events are mediated by enzymes, proteins that function in all cells. How the "information" is introduced that tells the cell what parts of the DNA to translate, and in what sequence, is only partly understood. With sufficient determination, it will no doubt be possible to "prove" that in some ultimate sense also this set of metabolic reactions is mediated by genes (DNA). We do know that the "message" encoded by DNA can be changed while it is translated into RNA and that proteins (hence further genes) are involved in all these processes. Therefore many genes must be involved in mediating every trait, irrespective of whether its transmission from one generation to the next follows the relatively simple pattern formally described by Mendel's laws or whether its transmission follows a more complicated pattern and hence is difficult to predict or describe.

Most traits geneticists study, and which biologists and other people think about, are of this complex sort. Many of them vary continuously, such as height. For these kinds of characters, called *quantitative traits,* reductionist explanations involving master molecules are clearly inappropriate.

Let me repeat: "Information" is not a good metaphor for what DNA contributes to protein synthesis. But even if it were, we would not be entitled to conclude that DNA controls or programs the many different ways in which proteins participate in the structure and functioning of organisms, not to speak of controlling or programming the complex characteristics of individuals and species.

How Genes Do and Do Not Function

Let us come back to genetics. I have said that the fact that a change in a gene registers as an observable difference in a character in no way implies that the gene controls or programs the character, only that it is one of the many components essential for the character to exist. I have now explained why this is so at the molecular level. But it is important to understand the genetic point because if we understand that the idea of genetic programming or control is based on a misinterpretation of how genes function, we can begin to understand why Mendel's laws can describe the transmission of only a few inherited traits and why some inherited traits follow Mendel's laws only some of the time.

Most traits do not follow Mendel's laws, and many are quantitative, not just present or absent. But often when a Mendelian pattern cannot be assigned, yet there is reason or inclination to believe the trait is inherited (which is conventionally taken to mean innate), the trait is said to be polygenic—that is, "controlled" by so many genes that the Mendelian yes/no pattern gets washed out. The assumption that many genes are involved when one will not explain the observation was criticized in the early days by Morgan (1909), who warned: "In the modern interpretation of Mendelism facts are being transformed into factors at a rapid rate. If one factor will not explain the facts, then two are invoked; if two prove insufficient, three will sometimes work out." Morgan and his colleagues (1915, p. 210) also pointed out "that a single factor may have several effects, and that a single character may depend on many factors."

All traits are polygenic in the sense that they are produced by many interacting, and often mutually regulating, processes that involve many enzymes and substrates, and therefore many genes. Occasionally, these relationships operate in such a way that changes in one particular gene are expressed as observable changes in the organism. But that does not mean that this gene causes the trait it affects. It is fortunate for geneticists that some of these changes follow Mendel's laws because these rare situations have made it possible to observe the transmission of this class of inherited traits among successive generations and to deduce information about the activities of genes.

One of the problems here is that the concept of the gene is used quite loosely. Genes are invoked to explain the origin of specific traits as well as of major structures and functions of organisms. They are called upon to program the orderly transformations during development and aging. And they are said to be decisive for slow, cumulative changes during evolution and species formation. Much of this speculation rests on assertions that are not based on observations. Often the appropriate experiments cannot be designed or even envisaged.

The identification of the gene as the double helix of DNA has increased the confusion. When a cell divides, the genes are duplicated, so that each daughter cell ends up with an identical set of genes. The beauty of the DNA structure is that one strand of the double helix serves as the template for the synthesis of its partner, so that DNA replication is quite straightforward. Unfortunately, this mode of reproduction of DNA has been accepted as a metaphor for the reproduction of likenesses between parents and offspring. But the replication of DNA is not a proper model for the inheritance of traits because the only thing that is copied in this way is DNA itself. Agata Mendel writes (p. 260):

What the complementary relations between thymine and adenine, and between guanine and cytosine explain, is that *when* a gene reproduces, an exact copy of the original is reproduced. If for one reason or another the gene is changed (which corresponds to a change in the base sequence), it is the changed form which will be copied during reproduction. Therefore what is in fact explained is that a *difference* in a gene will give rise to *differences* in its descendants [my translation].

I want to stress also that genes do not reproduce and DNA does not replicate itself, as they are sometimes said to do. Their reproduction or replication happens as part of the metabolism of living cells. Even if we understand in detail how DNA is synthesized and replicated in cells, this information cannot tell us how the replication of genes is translated into the transmission of traits from one generation to the next.

The diversity of living organisms, as well as the differentiation of cells and tissues during embryological development, results from an ongoing dialectic between the production of likeness and difference. By studying spontaneous and induced mutations, geneticists can learn about some of the intracellular correlates of difference (the genes); they cannot find out how likeness is produced.

That has been the task of embryologists. But embryology as well as genetics has been forced into the same reductionist paradigm. At present, molecular biologists try to study development by way of developmental genetics, which involves the production and analysis of so-called developmental mutants, organisms which exhibit an inherited defect that makes their development stop at one or another specific stage. This study of mutants can elucidate some of the sequences of steps that occur during normal development, but it cannot reveal what causes the individual steps or controls their spatiotemporal patterning (Wright, 1979). Agata Mendel jokes that molecular biologists tend to reason like a child who, because turning the knob on the television set makes the picture appear, concludes that the knob causes or programs the picture, and then goes on to the next, more absurd, step of trying to understand how television works by chemically analyzing the knob.

The reemergence of eugenics in the form of genetic engineering is the most recent stage in the drama of genes as determinants or causes. Its promise today is the same as was Muller's in 1913: to predict the characteristics of the unborn and modify them according to plan. Fletcher (1974) rhapsodizes:

> Insect societies are stable (stagnant?) for millenia because they are shaped by genetic transmission over which insects have no control. Unlike the locked-in bugs, men [sic] are at last able creatively to shape their own lives—genetically as well as culturally. . . . . When genetic therapy and surgery reach at last the goals already on their drawing boards we will have control over the quality of the infants to be born even *before* conception. . . . All of this means we have entered upon positive or *direct* eugenic control, and that we have surmounted the inadequacy and errors of negative control by selective mating.

By now it should be clear that I do not think this biocratic dream can be realized by manipulating genes because it exaggerates the control genes exert over metabolism and development. But the issue is not just whether genetic engineering and gene therapy will work in practice. The promise that people will be improved by manipulating our genes has an ideological impact, whether or not it can be done.

For more than a century people have been assured on scientific authority that the causes of our most serious personal and social problems reside within us. Now they are told that scientists soon will be able to cure all manner of ills—from sickle-cell anemia to manic depression and schizophrenia—by replacing "bad genes" with "good" ones. The fact that, under the direction of foreign genes, bacteria can be made to produce chemicals they would not produce normally (for example, human insulin or interferon) is cited as proof. But to engineer bacteria to make a specific product is not comparable to changing the genetic makeup of multicellular organisms or even of some of their cells and tissues. Bacteria do not provide proper models for the way genes participate in the functions of multicellular organisms, where patterning and control involve large numbers of interacting metabolites and pathways as well as interactions between cells and tissues and between the organism and its environment. Genes are only part of this story, and their roles are not sufficiently well understood to predict what will happen if one or another of them is changed, replaced, or even just moved from one position to another on the chromosomes. In fact, many molecular biologists believe that oncogenes—genes involved in cancerous growths—may be ordinary genes that are running amok because they somehow got into the wrong place.

The various attempts scientists have made to get genes they have inserted into multicellular organisms ("gene therapy") to do something have failed so far. Although molecular biologists have been able to make genes (DNA) induce bacteria or isolated cells, grown in tissue culture, to produce chemical products, when they have introduced genes into organized tissues or whole living organisms, the genes so far have been silent.

Sequencing the Human Genome

The reductionist belief in genes as causes has encouraged molecular biologists to undertake the gigantic project of trying to determine the base composition and sequence of the DNA (in other words, of the genes) in all twenty-three human chromosomes. In Part III I will look at some of the political and ethical questions this project and the exaggerated emphasis on genes raise. For now let us look at some scientific problems with the Human Genome Initiative.

We can begin by asking why anyone would want to undertake the herculean task of trying to identify and sequence the fifty to a hundred thousand genes estimated to make up the human genome and the approximately three billion nucleotide bases of which they are composed. The grandiose, reductionist reply offered by James Watson (1989) and a number of other molecular biologists is that this project will at last tell us what it means to be human—despite the fact that we seem to share 99 percent of our genes with the chimpanzees. (Or perhaps their very point is that when scientists have sequenced all the genes for humans and chimpanzees, they will at last be able to tell us apart.) A more modest claim is that having the complete DNA sequence will enable medical scientists to diagnose, and eventually cure, large numbers of genetically based diseases. Both claims are firmly grounded in the assumption that genes cause traits and therefore that the more genes we can characterize, the more we will know about how organisms function.

This is the wrong way to look at the situation. The fundamental question that needs to be answered is how the different cells in multicellular organisms come to assume quite different functions despite the fact that they all have the same genes. As I pointed out previously, the problem of differentiation lies at the heart of the way complex organisms develop from a fertilized egg, one single cell. Molecularly inclined embryologists try to explain this differentiation by saying, for example, that the cells in the different tissues become different from one another even though all of them contain the same complement of DNA because in different types of cells different genes get activated at different times. But that explanation begs the question by merely shifting the presumption of genetic "control" to a different level.

There is a historical reason why molecular biologists—the scientists who try to explain gene function in terms of the replication of DNA and its translation into the language of proteins—may not be conceptually in tune with the problem of differentiation. Their ways of thinking have developed out of working with bacteria and with viruses, called *bacteriophage* or simply *phage,* that can infect bacteria. Phage have only a few genes, and their job is to enter bacteria and get them to make more phage. Although bacteria are metabolically more complicated than phage, they do not need to cope with the problem of turning into cells that are as different from each other as nerve, muscle, liver, and kidney cells or of producing anatomical structures as different as a hand and an eye. Reductionist thinking, although limited, may yield useful answers when applied to phage or bacteria. But detailed information about our genes can answer only a limited range of questions about complicated creatures like ourselves.

This is not to say that some of these questions are not of scientific interest and not worth asking. But they could be asked much more conveniently with, say, fruit flies (drosophila). Scientists understand the genetics of fruit flies in considerable detail. Drosophila has only four pairs of chromosomes in its cells instead of our twenty-three pairs, and the locations of many genes on its chromosomes are known fairly accurately. What is more, the variability in the sequence of bases within genes has been studied extensively, so that scientists are beginning to have a sense of how different DNA sequences within a gene and different gene combinations are expressed, or not expressed, in different traits.

By studying the genetic map of fruit flies at the molecular level it should be possible to get some insight into the relationships between specific DNA sequences and specific anatomical or physiological characteristics of the whole organism. So, mapping the DNA sequence of drosophila, where we have so much genetic information and can do controlled genetic experiments, might make sense. But even here it would be more informative to look in detail at a few genes and compare their composition and the ways they function in many different individuals than to map all the genes of one or a few individuals.

The point of looking at a specific gene in a lot of different individuals is to try to differentiate coincidental correlations between base sequences and the appearance of a specific trait from those that are functionally related. For example, scientists have observed a number of different base sequences within a specific gene in drosophila without being able to see any difference in the trait with which the gene is associated. Yet some of these sequences also occur in individuals that, in fact, look different. Unless these scientists had analyzed a sufficiently large number of similar and different-looking individuals, they would have drawn the false conclusion that some of these base differences are responsible for the observable differences in the trait.

This sort of detailed analytical work cannot be done with people. Yet it would be necessary if one wants to know what significance to attach to any particular base, or indeed gene, sequence. But we have twenty-three pairs of chromosomes with many millions of genes on each, so the job of associating specific base sequences with specific traits is much more difficult than in drosophila. What is more important is that experimental work can be done ethically on only a limited number of people because there must be good reason to believe that it will benefit the individuals in question. One could try to get permission to collect a minute amount of tissue from huge numbers of people, say, by asking them to open their mouths and taking a tiny scrape off the inside of their cheeks. But because these people would not be well characterized genetically, and it would be ethically improper so to characterize them, the molecular information that could be obtained would be of little analytical value.

Anyway, the genome project is not intended to give us information about genetic diversity, which I am contending we would need in order to know what significance to attach to any specific DNA sequence. It is intended to produce the complete DNA sequence of the twenty-three chromosomes for a human "prototype," which would be a composite of chromosomal regions obtained from the cells and tissues of different people.

The scientific significance of this laborious and expensive effort is as questionable as was the scientific significance of putting a man on the moon. But it has a similar, heroic appeal. The problem is that, quite aside from the waste of resources—both money and scientific personnel—the human genome project will have unfortunate practical and ideological consequences because it will increase the mythic importance our culture assigns to genes and genetic inheritance.

7

Have Only Men Evolved?

With the dawn of scientific investigation it might have been hoped that the prejudices resulting from lower conditions of human society would disappear, and that in their stead would be set forth not only facts, but deductions from facts, better suited to the dawn of an intellectual age. . . .

The ability, however, to collect facts, and the power to generalize and draw conclusions from them, avail little, when brought into direct opposition to deeply rooted prejudices.

—Eliza Burt Gamble,
The Evolution of Women (1894)

Scientific explanations have repeatedly run counter to beliefs held dear by powerful segments of the society. Organized religion, for example, has its own explanations of how nature works. For this reason, scientists are sometimes portrayed as lone heroes swimming against the social stream, and Darwin's theories of evolution and human descent are frequently used to illustrate this point. But Darwinism has wide areas of congruence with the social and political ideology of nineteenth-century Britain and with Victorian precepts of morality, particularly as regards the relationships between the sexes. And, as we shall see in chapter 8, the same Victorian notions still dominate contemporary biological thinking about sex differences and sex roles.

Darwin's Evolutionary Theory

It is interesting that the idea that Darwin was swimming against the stream has prevailed in spite of the fact that historians have shown that his thinking fitted squarely into the historical and social perspective of his time. Darwin so clearly and admittedly was drawing together strands that had been developing over long periods that the questions of why he was the one to produce the synthesis and of why he did so just then have clamored for answers. Therefore, the social origins of the Darwinian synthesis have been probed by numerous scientists and historians.

A belief that all living forms are related and that there also are deep connections between the living and nonliving has existed during much of recorded human history. Through the animism of tribal cultures, which endows everyone and everything with a common spirit; through elaborate expressions of the unity of living forms in some Far Eastern and Native American belief systems; and through Aristotelian notions of connectedness runs the theme of a web of life that includes humans among its many strands.

The Judaeo-Christian world-view has been exceptional—and I would say flawed—in setting man (and I mean the male of the species) apart from the rest of nature by making him the namer and ruler of all life. The biblical myth of the creation gave rise to the separate and unchanging species which Carolus Linnaeus, that second Adam, named and classified in the eighteenth century. But even though Linnaeus began by accepting the belief that all existing species had been created by Jehovah during that one week long ago, he had his doubts about their immutability by the time he had identified more than four thousand of them: Some species appeared to be closely related, others seemed clearly transitional. Yet, as Loren Eiseley (1961, p. 24) has pointed out, it is important to realize that "until the scientific idea of 'species' acquired form and distinctness there could be no dogma of 'special' creation in the modern sense. This form and distinctness it did not possess until the naturalists of the seventeenth century began to substitute exactness of definition for the previous vague characterizations of the objects of nature." And he continues: "It was Linnaeus with his proclamation that species were absolutely fixed since the beginning who intensified the theological trend. . . . Science, in its desire for classification and order, . . . found itself satisfactorily allied with a Christian dogma whose refinements it had contributed to produce."

Did species exist before they were invented by scientists with their predilection for classification and naming? And did the new science, by concentrating on differences which could be used to tell things apart, devalue the similarities that tie them together? Certainly the Linnaean system succeeded in congealing into a relatively static form what had been a more fluid and graded world that allowed for change and hence for a measure of historicity.

The hundred years that separate Linnaeus from Darwin saw the development of historical geology by Charles Lyell and an incipient effort to fit the increasing number of fossils that were being uncovered into the earth's newly discovered history. By the time Darwin came along, it was clear to many people that the earth and its creatures had histories. There were fossil series of snails; some fossils were known to be very old, yet looked for all the world like present-day forms; others had no like descendants and had become extinct. Jean de Lamarck, who like Linnaeus began by believing in the fixity of species, by 1800 had formulated a theory of evolution that involved a slow historical process, which he assumed to have taken a very long time.

Possibly one reason the theory of evolution arose in Western, rather than Eastern, science was the fact that the descriptions of fossils and living forms showed so many close relationships. These relationships made the orthodox biblical view of the special creation of each and every species untenable, so the question of how living forms are related to one another pressed for an answer. The Eastern philosophies that accept connectedness and relatedness as givens did not need to confront this question with the same urgency. In other words, where evidences of evolutionary change did not raise fundamental contradictions and questions, evolutionary theory did not need to be invented to reconcile and answer them. However one, and perhaps the most, important difference between Western evolutionary thinking and Eastern ideas of organismic unity lies in the materialistic and historical elements, which are the earmark of Western evolutionism as formulated by Darwin.

Although most of the elements of Darwinian evolutionary theory existed for at least a hundred years before Darwin, he knit them into a consistent theory that was in line with the mainstream thinking of his time. William Irvine (1972, p. 98) writes:

The similar fortunes of liberalism and natural selection are significant.
Darwin's matter was as English as his method. Terrestrial history turned
out to be strangely like Victorian history writ large. Bertrand Russell
and others have remarked that Darwin's theory was mainly "an extension to the animal and vegetable world of laissez faire economics." As a
matter of fact, the economic conceptions of utility, pressure of population, marginal fertility, barriers in restraint of trade, the division of
labor, progress and adjustment by competition, and the spread of technological improvements can all be paralleled in the *The Origin of Species*.
But so, alas, can some of the doctrines of English political conservatism.
In revealing the importance of time and the hereditary past, in empha-
sizing the persistence of vestigial structures, the minuteness of variations
and the slowness of evolution, Darwin was adding Hooker and Burke to
Bentham and Adam Smith. The constitution of the universe exhibited
many of the virtues of the English constitution.

One of the first to comment on this congruence was Karl
Marx, who wrote Friedrich Engels in 1862, three years after the
publication of *The Origin of Species* (quoted in Sahlins, 1976,
p. 102):

> It is remarkable how Darwin recognizes among beasts and plants his En-
> glish society with its division of labour, competition, opening up new
> markets, "inventions," and the Malthusian "struggle for existence." It is
> Hobbes' "bellum omnium contra omnes" [war of all against all], and one
> is reminded of Hegel's *Phenomenology*, where civil society is described as
> a "spiritual animal kingdom," while in Darwin the animal kingdom fig-
> ures as civil society.

A similar passage appears in a letter by Engels (Sahlins, 1976,
p. 103):

> The whole Darwinist teaching of the struggle for existence is simply a
> transference from society to living nature of Hobbes's doctrine of "bellum
> omnium contra omnes" and of the bourgeois-economic doctrine of com-
> petition together with Malthus's theory of population. When this con-
> jurer's trick has been performed . . . the same theories are transferred
> back again from organic nature into history and now it is claimed that
> their validity as eternal laws of human society has been proved.

The very fact that essentially the same mechanism of evolution
through natural selection was postulated independently by
Darwin and by his contemporary, Alfred Russel Wallace, shows
that the basic ideas were in the air, which is not to deny that it
took genius to give them logical and convincing form.

Darwin's *The Origin of the Species by Means of Natural Selection*, published in 1859, accepted the fact of evolution and undertook to explain how it could have come about. Darwin had amassed large quantities of data to show that historical change had taken place, both from the fossil record and from his observations as a naturalist on the *Beagle*. He pondered why some forms had become extinct and others had survived to generate new and different forms. The watchword of evolution seemed to be: "be fruitful and modify," which bore a striking resemblance to the ways of animal and plant breeders. Darwin corresponded with many breeders and himself began to breed pigeons. He was impressed by the way in which breeders, through careful selection, could use even minor variations to elicit major differences and was searching for the analog in nature to the breeders' techniques of selecting favorable variants. A prepared mind therefore encountered Malthus's *Essay on the Principles of Population* (1798). In his *Autobiography*, Darwin writes (Francis Darwin, 1958, pp. 42–43):

> In October 1838, that is, fifteen months after I had begun my systematic enquiry, I happened to read for amusement Malthus on *Population*, and being well prepared to appreciate the struggle for existence which everywhere goes on from long-continued observation of the habits of animals and plants, it at once struck me that under these circumstances favorable variations would tend to be preserved and unfavorable ones to be destroyed. The result of this would be the formation of new species. Here, then, I had at last got a theory by which to work.

Wallace also acknowledged being led to his theory by reading Malthus (quoted in Francis Darwin, 1958, pp. 200–201):

> The most interesting coincidence in the matter, I think, is, that I, *as well as Darwin*, was led to the theory itself through Malthus. . . . It suddenly flashed upon me that all animals are necessarily thus kept down—"the struggle for existence"—while *variations*, on which I was always thinking, must necessarily often be *beneficial*, and would then cause those varieties to increase while the injurious variations diminished [Wallace's italics].

Both, therefore, saw in Malthus's struggle for existence the working of a natural law which effected what Spencer had called "the survival of the fittest."

The three principal ingredients of Darwin's theory of evolution are endless variation, natural selection from among the variants, and the resulting survival of the fittest. Given the looseness of many of his arguments—he credited himself with being an expert wriggler—it is surprising that his explanation has found such wide acceptance. One reason probably lies in the fact that Darwin's theory was historical and materialistic, two esteemed characteristics; another, perhaps, in its intrinsic optimism—its notion of progressive development of species, one from another—which fit well into the meritocratic ideology encouraged by the early successes of British mercantilism, industrial capitalism, and imperialism.

Darwin's interpretation of the history of life on earth not only fit well into the social doctrines of nineteenth-century liberalism and individualism but was used, in turn, to support them by rendering them aspects of natural law. Spencer is usually credited with having brought Darwinism into social theory. The body of ideas, which came to be known as Social Darwinism, gained wide acceptance in Britain and the United States in the latter part of the nineteenth and on into the twentieth century. For example, John D. Rockefeller proclaimed in a Sunday school address (quoted in Hofstadter, 1955, p. 45): "The growth of a large business is merely the survival of the fittest. . . . The American Beauty rose can be produced in the splendor and fragrance which bring cheer to its beholder only by sacrificing the early buds which grow up around it. This is not an evil tendency in business. It is merely the working-out of a law of nature and a law of God." The circle was therefore complete: Darwin consciously borrowed from social theorists such as Malthus and Spencer some of the basic concepts of evolutionary theory. Spencer and others promptly used Darwinism to reinforce these very social theories and in the process bestowed upon them the force of natural law.

Sexual Selection

It is essential to expand the foregoing analysis of the mutual influences of Darwinism and nineteenth-century social doctrine by looking critically at the Victorian picture Darwin painted of the relations between the sexes and of the roles males and females play in the evolution of animals and humans. The ethnocentric bias of Darwinism has been widely acknowledged, but its blatant sexism—or, more correctly, androcentrism (male-centeredness)—although noted by a few nineteenth-century feminists, went largely unexplored until the 1970s, when feminist scientists and historians became interested in Darwin. Within Darwin's lifetime, feminists such as Antoinette Brown Blackwell and Eliza Burt Gamble called attention to the obvious male bias pervading his arguments. But these women did not have Darwin's or Spencer's professional status or scientific experience; nor indeed could they have, given their limited opportunities for education, travel, and participation in the affairs of the world. Their books were hardly acknowledged or discussed by professionals, and they have been excluded from the record. It is important to expose Darwin's androcentrism not only for historical reasons but because it remains an integral and unquestioned part of contemporary biological theories.

Early in *The Origin of Species,* Darwin ([1859] n.d.) defines what he calls *sexual selection* as one mechanism by which evolution operates. The Victorian and androcentric biases are obvious (p. 69):

> This form of selection depends, not on a struggle for existence in relation to other organic beings or to external conditions, but on a struggle of individuals of one sex, generally males, for the possession of the other sex. . . . Generally, the most vigorous males, those which are best fitted for their places in nature, will leave most progeny. But in many cases, victory depends not so much on general vigour, as on having special weapons confined to the male sex.

The Victorian picture of the active male and the passive female becomes even more explicit later in the same paragraph: "The males of certain hymenopterous insects [bees, wasps, ants] have been frequently seen by that inimitable observer, M. Fabre, fighting for a particular female who sits by, an apparently unconcerned beholder of the struggle, and then retires with the conqueror."

Darwin's anthropomorphizing continues, as it develops that many male birds "perform strange antics before the females, which standing by as spectators, at last choose the most attractive partner." However, he worries that whereas this might be a reasonable way to explain the behavior of peahens and female birds of paradise, whose consorts anyone can admire, "it is doubtful whether [the tuft of hair on the breast of the wild turkey cock] can be ornamental in the eyes of the female bird." Hence Darwin ends this brief discussion by saying that he "would not wish to attribute all sexual differences to this agency" (p. 70).

Some might argue in defense of Darwin that bees (or birds or what have you) do act that way. But the language Darwin uses to describe these behaviors disqualifies him as an objective observer. His animals are cast into roles from a Victorian script. And whereas no one can claim to have solved the important methodological question of how to disembarrass oneself of one's anthropocentric and cultural biases when observing animal behavior, surely one must begin by trying.

After the publication of *The Origin of Species*, Darwin continued to think about sexual selection, and, in 1871, he published *The Descent of Man and Selection in Relation to Sex*, a book in which he describes in more detail than in *The Origin of Species* how sexual selection operates in the evolution of animals and humans.

In the aftermath of the outcry *The Descent* raised among religious fundamentalists, much has been made of the fact that Darwin threatened the special place "man" was assigned by the Bible and treated him as though he were just another kind of animal. But he did nothing of the sort. The Darwinian synthesis did not end anthropocentrism or androcentrism in biology. On the contrary, Darwin made them part of biology by presenting as "facts of nature" interpretations of animal behavior that reflect the social and moral outlook of his time.

In a sense, anthropocentrism is implicit in the fact that we humans have named, catalogued, and categorized the world around us, including ourselves. Whether we stress our upright stance, our opposable thumbs, our brain, or our language, to ourselves we are creatures apart and different from all others. But the scientific view of ourselves is also profoundly androcentric. *The Descent of Man* is quite literally *his* journey. Elaine Morgan (1973, pp. 3–4) rightly says:

It's just as hard for man to break the habit of thinking of himself as central to the species as it was to break the habit of thinking of himself as
central to the universe. He sees himself quite unconsciously as the main
line of evolution, with a female satellite revolving around him as the
moon revolves around the earth. This not only causes him to overlook
valuable clues to our ancestry, but sometimes leads him into making
statements that are arrant and demonstrable nonsense. . . . Most of the
books forget about [females] for most of the time. They drag her on
stage rather suddenly for the obligatory chapter on Sex and Reproduction, and then say: "All right, love, you can go now," while they get on
with the real meaty stuff about the Mighty Hunter with his lovely new
weapons and his lovely new straight legs racing across the Pleistocene
plains. Any modifications of her morphology are taken to be imitations
of the Hunter's evolution, or else designed solely for his delectation.

To expose the Victorian roots of post-Darwinian thinking
about human evolution, we must start by looking at Darwin's
ideas about sexual selection in *The Descent,* where he begins the
chapter entitled "Principles of Sexual Selection" by setting the
stage for the active, pursuing male ([1871] n.d., p. 567):

With animals which have their sexes separated, the males necessarily
differ from the females in their organs of reproduction; and these are
the primary sexual characters. But the sexes differ in what Hunter has
called secondary sexual characters, which are not directly connected
with the act of reproduction; for instance, the male possesses certain
organs of sense or locomotion, of which the female is quite destitute, or
has them more highly developed, in order that he may readily find or
reach her; or again the male has special organs of prehension for holding her securely.

Moreover, we soon learn that "in order that the males should
seek efficiently, it would be necessary that they should be endowed with strong passions; and the acquirement of such passions would naturally follow from the more eager leaving a
larger number of offspring than the less eager" (p. 580).

But Darwin is worried that among some animals, males and
females do not appear to be all that different, as though "a
double process of selection has been carried on; that the males
have selected the more attractive females, and the latter the
more attractive males. . . . But from what we know of the habits
of animals, this view is hardly probable, for the male is gener-

ally eager to pair with any female" (p. 582). Make no mistake, wherever you look among animals, eagerly promiscuous males are pursuing females, who peer from behind languidly drooping eyelids to discern the strongest and handsomest. Does it not sound like the wish-fulfillment dream of a proper Victorian gentleman?

This is not the place to discuss Darwin's long treatise in detail. Therefore, let this brief look at animals suffice as background for his section on "Sexual Selection in Relation to Man." Again we can start on the first page: "Man is more courageous, pugnacious and energetic than woman, and has more inventive genius" (p. 867). Among "savages," fierce, bold men are constantly battling each other for the possession of women, and this battle has affected the secondary sexual characteristics of both. Darwin grants that there is some disagreement whether there are "inherent differences" between men and women, but suggests that by analogy with lower animals it is "at least probable." In fact, "woman seems to differ from man in mental disposition, chiefly in her greater tenderness and less selfishness," for "man is the rival of other men; he delights in competition, and this leads to ambition which passes too easily into selfishness. These latter qualities seem to be his natural and unfortunate birthright" (p. 873).

This difference in qualities might seem to make women better than men after all, but not so (pp. 873–874):

> The chief distinction in the intellectual powers of the two sexes is shown by man's attaining to a higher eminence, in whatever he takes up, than can woman—whether requiring deep thought, reason, or imagination, or merely the use of the senses and hands. If two lists were made of the most eminent men and women in poetry, painting, sculpture, music (inclusive both of composition and performance), history, science, and philosophy, with half-a-dozen names under each subject, the two lists would not bear comparison. We may also infer . . . that if men are capable of a decided preeminence over women in many subjects, the average of mental power in man must be above that of woman. . . . [Men have had] to defend their females, as well as their young, from enemies of all kinds, and to hunt for their joint subsistence. But to avoid enemies or to attack them with success, to capture wild animals and to fashion weapons requires the aid of the higher mental faculties, namely, observation, reason, invention, or imagination. These various faculties will thus have been continually put to the test and selected during manhood.

"Thus," the discussion ends, "man has ultimately become superior to woman," and it is a good thing that men pass on their characteristics to their daughters as well as to their sons, "otherwise it is probable that man would have become as superior in mental endowment to woman, as the peacock is in ornamental plumage to the peahen" (p. 874).

So here it is in a nutshell: Men's mental and physical qualities were constantly improved through competition for women and hunting, while women's minds would have become vestigial if it were not for the fortunate circumstance that in each generation daughters inherit brains from their fathers.

Another example of Darwin's acceptance of the conventional mores of his time is his interpretation of the evolution of marriage and monogamy (p. 895):

> It seems probable that the habit of marriage, in any strict sense of the word, has been gradually developed; and that almost promiscuous or very loose intercourse was once very common throughout the world. Nevertheless, from the strength of the feeling of jealousy all through the animal kingdom, as well as from the analogy of lower animals . . . I cannot believe that absolutely promiscuous intercourse prevailed in times past.

Note the moralistic tone; and how does Darwin know that strong feelings of jealousy exist "all through the animal kingdom"?

For comparison, it is interesting to look at Engels, who, working largely from the same early anthropological sources as Darwin, had this to say (in Leacock, 1972, p. 138):

> As our whole presentation has shown, the progress which manifests itself in these successive forms [from group marriage to pairing marriage to what he refers to as "monogamy supplemented by adultery and prostitution"] is connected with the peculiarity that women, but not men, are increasingly deprived of the sexual freedom of group marriage. In fact, for men group marriage actually still exists even to this day. What for the woman is a crime entailing grave legal and social consequences is considered honorable in a man or, at the worse, a slight moral blemish which he cheerfully bears. . . . Monogamy arose from the concentration of considerable wealth in the hands of a single individual—a man—and from the need to bequeath this wealth to the children of that man and of no other. For this purpose, the monogamy of the woman was required, not that of the man, so this monogamy of the woman did not in any way interfere with open or concealed polygamy on the part of the man.

Clearly, Engels did not accept the Victorian code of behavior as our natural biological heritage.

Other Forms of Scientific Sexism

The theory of sexual selection went into a decline during the first half of this century, as efforts to verify some of Darwin's examples showed that many of the features he had thought were related to success in mating could not be legitimately regarded that way. But it has lately regained respectability, and contemporary discussions of reproductive fitness often cite examples of sexual selection. Therefore, before we go on to discuss human evolution, it is helpful to look at contemporary views of sexual selection and sex roles among animals (and even plants).

Let us start with a lowly alga that one might think impossible to stereotype by sex. Wolfgang Wickler (1973, p. 23), an ethologist at the University of Munich, writes in his book on sexual behavior patterns (a topic which Konrad Lorenz tells us in the Introduction is crucial in deciding which sexual behaviors to consider healthy and which diseased [sic]): "Even among very simple organisms such as algae, which have threadlike rows of cells one behind the other, one can observe that during copulation the cells of one thread act as males with regard to the cells of a second thread, but as females with regard to the cells of a third thread. The mark of male behavior is that the cell actively crawls or swims over to the other; the female cell remains passive." The circle is simple to construct: One starts with the Victorian stereotype of the active male and the passive female, then looks at animals, algae, bacteria, people, and calls all passive behavior feminine, all active or goal-oriented behavior masculine. And it works! The Victorian stereotype is biologically determined: Even algae behave that way.

But let us see what Wickler has to say about Rocky Mountain bighorn sheep, in which the sexes cannot be distinguished on sight. He finds it "curious" "that between the extremes of rams over eight years old and lambs less than a year old one finds every possible transition in age, but no other differences whatever. . . . The bodily form, the structure of the horns, and the color of the coat are the same for both sexes." Now note: "The typical female behavior is absent from this pattern" (p. 28). Typical of what? Obviously not of bighorn sheep. In fact we are told that "even males often cannot recognize a female"; indeed, "the females are only of interest to the males during rutting season" (p. 29). How does he know that the males do not recognize the females? Maybe these sheep are so weird that most of the time they relate to a female as though she were just another sheep and whistle at her (my free translation of taking an "interest") only when it is a question of mating.

But let us get at last to how the females behave. That is astonishing, for it turns out "that *both* sexes play two roles, either that of the male or that of the young male. Outside the rutting season the females behave like young males, during the rutting season like aggressive older males" (Wickler's italics). In fact: "There is a line of development leading from the lamb to the high ranking ram, and the female animals . . . behave exactly as though they were in fact males . . . whose development was retarded. . . . We can say that the only fully developed mountain sheep are the powerful rams." At last the androcentric paradigm is in the open: Females are always measured against the standard of the male. Sometimes they are like young males, sometimes like older ones; but never do they reach what Wickler calls "the final stage of fully mature physical structure and behavior possible to this species" (pp. 29–30). That, in his view, is reserved for the rams.

Wickler bases this discussion on observations by Valerius Geist (1971), whose book *Mountain Sheep* contains many examples of how androcentric biases can color observations as well as interpretations and restrict the imagination to stereotypes. One of the most interesting is the following (p. 190):

> Matched rams, usually strangers, begin to treat each other like females and clash until one acts like a female. This is the loser in the fight. The rams confront each other with displays, kick each other, threat jump, and clash till one turns and accepts the kicks, displays, and occasional mounts of the larger without aggressive displays. The loser is not chased away. The point of the fight is not to kill, maim, or even drive the rival off, but to treat him like a female.

This description would be quite different if the interaction were interpreted as something other than a fight, say as a homosexual encounter, a game, or a ritual dance. The fact is that it contains none of the elements we commonly associate with fighting. Yet because Geist casts it into the imagery of heterosexuality and aggression, it becomes perplexing.

There would be no reason to discuss these examples if their treatments of sex differences or of male and female behavior were exceptional. But they are in the mainstream of contemporary sociobiology, ethology, and evolutionary biology. A book that has been a standard reference is George Williams's (1975) *Sex and Evolution*. It abounds in blatantly biased statements that describe as "careful" and "enlightened" research reports that support the androcentric paradigm, and as questionable or erroneous those that contradict it. Masculinity and femininity are discussed with reference to the behavior of pipefish and seahorses; and cichlids and catfish are judged downright abnormal because both sexes guard the young.

For present purposes it is sufficient to discuss a few points raised in the chapter Williams entitles "Why Are Males Masculine and Females Feminine and, Occasionally, Vice-Versa?" The very title gives one pause, for if the words *masculine* and *feminine* do not mean of or pertaining to males and females, what do they mean in a scientific context? So let us read. On the first page we find: "Males of the more familiar higher animals take less of an interest in the young. In courtship they take a more active role, are less discriminating in choice of mates, more inclined toward promiscuity and polygamy, and more contentious among themselves." We are back with Darwin. The data are as flimsy as ever, but doesn't it sound like a description of the families on your block?

The important question is who are these "more familiar higher animals"? Is their behavior typical, or are we familiar with them because, for over a century, androcentric biologists have paid disproportionate attention to animals whose behavior resembles those human social traits that they would like to interpret as biologically determined and hence out of our control?

Williams's generalization gives rise to the paradox that becomes his chief theoretical problem: "Why, if each individual is maximizing its own genetic survival, should the female be less anxious to have her eggs fertilized than a male is to fertilize them, and why should the young be of greater interest to one than to the other?" Let me translate this sentence for the benefit of those unfamiliar with current evolutionary theory. The first point is that an individual's fitness is measured by the number of her or his offspring that survive to reproductive age. The phrase "the survival of the fittest" therefore signifies the fact that evolutionary history is the sum of the stories of those who leave the greatest numbers of descendants. What is meant by each individual "maximizing its own genetic survival" is that everyone tries to leave as many viable offspring as possible. (Note the implication of conscious intent. Such intent is not exhibited by the increasing number of humans who intentionally limit the numbers of their offspring. Nor is one justified in ascribing intention to other animals.)

One might therefore think that in animals in which each parent contributes half of each offspring's genes, females and males would exert themselves equally to maximize the number of offspring. However, we know that, according to the patriarchal paradigm, males are active in courtship, whereas females wait passively. This is what Williams means by females being "less anxious" to procreate than males. And of course we also know that "normally" females have a disproportionate share in the care of their young.

So why these asymmetries? The explanation: "The *essential* difference between the sexes is that females produce large immobile gametes and males produce small mobile ones" (p. 124; my italics). This difference determines their "different optimal strategies." So if you have wondered why men are promiscuous and women faithfully stay home and care for the babies, the reason is that males "can quickly replace wasted gametes and be ready for another mate," whereas females "can not so readily replace a mass of yolky eggs or find a substitute father for an expected litter." Therefore females must "show a much greater degree of caution" in the choice of a mate than males. Although these descriptions fit only some animal species that procreate sexually and are ceasing to fit human domestic arrangements in many portions of the globe, they do fit the patriarchal model of

the household. Clearly, androcentric biology is busy as ever trying to provide biological "reasons" for our social arrangements.
   In 1948, Ruth Herschberger wrote a delightfully funny book
called *Adam's Rib,* in which she spoofed androcentric myths regarding sex differences. In the chapter entitled "Society Writes
Biology," she translates the patriarchal scenario of the dauntless
voyage of the active, agile sperm toward the passively receptive,
sessile egg into an improvised "matriarchal" account. In it the
large, competent egg plays the central role, and we can feel
only pity for the many millions of minuscule, fragile sperm,
most of which are too feeble to make it to fertilization. Just look
at the language we use to describe fertilization. We say that a
sperm fertilizes an egg (active verb) and that the egg is fertilized
(passive verb). This grammar does not describe the biological
reality, which is that the two cells fuse, a process in which both
participate actively. It reflects the ideology of gender relations,
in which males pursue and females yield.
   The recent resurrection of the theory of sexual selection and
the ascription of asymmetry to the "parental investments" of
males and females are probably not unrelated to the rebirth of
the women's movement. We should remember that Darwin's
theory of sexual selection was put forward in the midst of the
first wave of feminism. It seems that when women threaten to
enter as equals into the world of affairs, androcentric scientists
rally to point out that women's natural place is in the home.

The Evolution of Man

Darwin's sexual stereotypes are doing well also in the contemporary literature on human evolution. In this field facts are few
and specimens are separated often by hundreds of thousands
of years, so that maximum leeway exists for investigator bias.
Until recently, almost all the investigators have been men. It
should therefore come as no surprise that what has emerged is
the familiar picture of Man the Toolmaker. This pattern ex-
tends so far that when skull fragments estimated to be two hundred fifty thousand years old turned up among the stone tools
in the gravel beds of the Thames at Swanscombe and paleon-

tologists decided that they are probably those of a woman, we read: "The Swanscombe woman, or her husband, was a maker of hand axes" (Howells, 1973, p. 88). Imagine the reverse: "The Swanscombe man, or his wife, was a maker of axes." The implication is that if there were tools, the Swanscombe woman could not have made them. But even apes make tools. Why not women?

Actually, the idea that making and using tools were the main driving forces in evolution has been modified since paleontological finds and field observations have shown that apes both use and fashion tools. Now the emphasis is on the human use of tools as weapons for hunting. This brings us to Man the Hunter, who had to invent not only tools but also the social organization that allowed him to hunt big animals. He also had to roam great distances and learn to cope with many and varied circumstances. We are told that this entire constellation of factors stimulated the astonishing and relatively rapid development of his brain, which came to distinguish him from his ape cousins. For example, Kenneth Oakley (1972, p. 81) writes:

> Men who made tools of the standard type . . . must have been capable of forming in their minds images of the ends to which they laboured. Human culture in all its diversity is the outcome of this capacity for conceptual thinking, but the leading factors in its development are tradition coupled with invention. The primitive hunter made an implement in a particular fashion largely because as a child he watched his father at work or because he copied the work of a hunter in a neighbouring tribe. The standard hand-axe was not conceived by any one individual *ab initio*, but was the result of exceptional individuals in successive generations not only copying but occasionally improving on the work of their predecessors. As a result of the co-operative hunting, migrations and rudimentary forms of barter, the traditions of different groups of primitive hunters sometimes became blended.

It seems a remarkable feat of clairvoyance to see in such detail what happened some two hundred fifty thousand years in prehistory, complete with the little boy and his little stone chipping set just like daddy's big one.

It is hard to know what reality lurks behind such reconstructions of Man Evolving. Since the time when we and the apes diverged some five million years ago, the main features of human evolution that one can read from paleontological finds are the upright stance, reduction in the size of the teeth, and increase in brain size. But finds are few and far between both in space and time until we reach the Neanderthals, some seventy to forty thousand years ago—a jaw or skull, teeth, pelvic bones, and often only fragments of them. From such bits of evidence as these come the pictures and statues we have all seen of that line of increasingly straight and upright, and decreasingly hairy and apelike, men marching in single file behind *Homo sapiens,* carrying their clubs, stones, or axes; or that other one of a group of beetle-browed and bearded hunters bending over the large slain animal they have brought into camp, while over by the side long-haired, broad-bottomed females nurse infants at their pendulous breasts.

Impelled, I suppose, by feminist critiques of the evolution of Man the Hunter, such as those of Nancy Tanner and Adrienne Zihlman (Tanner, 1981; Tanner and Zihlman, 1976), anthropologists have begun to take note of contributions women must have made in early human societies, but the stereotyping continues. For example, William Howells (1973), who grants that such criticisms are valid, nonetheless assumes a "classic division of labor between the sexes" and states as fact that stone-age men roamed great distances "on behalf of the whole economic group, while the women were restricted to within the radius of a fraction of a day's walk from camp" (p. 133). Needless to say, he does not *know* any of this.

One can equally assume that the responsibilities for providing food and nurturing young were widely dispersed through the group because it needed to cooperate and devise many and varied strategies for survival. Nor is it obvious why tasks need have been differentiated by sex. It makes sense that the gatherers would have known how to hunt the animals they came across; that the hunters gathered when there was nothing to catch; and that men and women did some of each, although both of them probably did a great deal more gathering than hunting. After all, the important thing was to get food, not to define sex roles. Bearing and tending the young do not necessitate a sedentary way of life among foragers and nomadic peoples right to the present, and both gathering and hunting must have required moving over large areas in order to find sufficient food. Hewing close to home probably accompanied the transition to cultivation, which introduced the necessity to stay put for planting, although only until the harvest. Without fertilizers and crop rotation, frequent moves were probably essential parts of early farming.

Being sedentary ourselves, we tend to assume that our foreparents heaved a sigh of relief when they invented agriculture and could at last stop roaming. But foragers and other people who move with their food still exist. And what has been called the agricultural "revolution" probably took considerably longer than all of recorded history. During this time, presumably some people settled down while others kept moving, and yet others did some of each, depending on place and season. We have developed a fantastically limited and stereotypic picture of ways of life that evolved over many tens of thousands of years and no doubt varied in lots of ways that we do not even imagine. It is true that by historic times, which are only an instant in our evolutionary history, there were agricultural settlements, including a few towns that numbered hundreds and even thousands of inhabitants. By that time labor was to some extent divided by sex, although antropologists have shown that right to the present the division can be different in different places. There are economic and social reasons for the various delineations of sex roles. We presume too much when we try to read them in the scant record of our prehistoric past.

Nor are we going to learn them by observing our nearest living relatives among the apes and monkeys, as some biologists and anthropologists are trying to do. For one thing, different species of primates vary widely in the extent to which the sexes differ in both anatomy and social behavior, so that one can find examples of almost any kind of behavior one is looking for by picking the appropriate animal. For another, most scientists find it convenient to forget that present-day apes and monkeys have had as long an evolutionary history as human beings have had, since the time we went our separate ways. There is no theoretical reason why their behavior should tell us more about our ancestry than our behavior tells us about theirs. But just as in the androcentric paradigm men evolved while women cheered from the bleachers, so in the anthropocentric one humans evolved while the apes watched from the trees.

Although the current literature on human evolution says much about behaviors traditionally associated with men, it says little about the role of language and sometimes even associates the evolution of language with tool use and hunting—two supposedly masculine characteristics. But such a link is unlikely because the evolution of language probably went with changes in the structure of the face, larynx, and brain, all slow processes. Tool use and hunting, however, are cultural characteristics that can evolve much more quickly than anatomical or physiological ones. It is likely that the elaborate use of tools and the social arrangements that go with hunting and gathering developed in part as a consequence of expanded human capacities and needs that derive from our ability to communicate through language.

The evolution of speech may have been a powerful force directing our biological, cultural, and social evolution. So it is surprising that its significance has been largely ignored by biologists. But, of course, it does not fit into the andocentric paradigm. No one has ever claimed women cannot talk. Thus, if men are the vanguard of evolution, humans must have evolved through stereotypically male behaviors, such as competition, tool use, and hunting.

8

believe that the "lower" levels, such as atoms and molecules, are
more "basic" and have intrinsically greater explanatory pow
As we have seen, molecular biologists have described genes
tive, which has been organized to determine the composition
and sequence of all the genes on the human chromosomes, for
its faith in the superior explanatory power of the "lower" levels,
its reductionism.

Human Nature

 THE ambiguity of the term *biology* is at the heart of questions
about what scientists do when they try to examine nature. We
use the term to denote what scientists tell us about organisms
and also the living experience itself. When I speak of "my biology," I usually mean the ways I experience my biological functions, not what scientists tell me about them. I can also use the
word to denote the scientific discipline, as in "I am studying biology." This confusion about what we mean by biology reflects
ambiguities about the connections between scientific descriptions of the world and the phenomena scientists try to describe,
such as those I discussed in Part I. We need to be aware of these
ambiguities when we think about human nature because it does
not describe real people. It is an abstraction, or reification, a
normative concept that incarnates (in the literal sense of en-
veloping in flesh) historically based beliefs about how people
should behave.

  Biologists' descriptions of human nature are imbedded in the
ways we study organisms. Most scientists accept the notion that
nature can best be described in levels of organization that extend from ultimate particles, via atoms and molecules, to cells,
tissues, and organs, to organisms considered individually, and
then on to groups of organisms (that is, societies). Biology is
concerned with the range of levels from atoms and molecules
through organisms, and also with the relationships between different organisms over time (evolution) and space (animal behavior and ecology). Some biologists study organisms by taking
them apart, while others observe whole organisms in the laboratory or the field. Yet, in the current view, these levels do not
have equal prestige and are not credited with equal explanatory power. Most biologists, as well as chemists and physicists,

believe that the "lower" levels, such as atoms and molecules, are more "basic" and have intrinsically greater explanatory power. As we have seen, molecular biologists have described genes (which are molecules) as keys to the "secret of life" or blueprints of the organism. I have criticized the Human Genome Initiative, which has been organized to determine the composition and sequence of all the genes on the human chromosomes, for its faith in the superior explanatory power of the "lower" levels, its reductionism.

Reductionism operates across levels. Reductionists believe that by studying individuals we will come to understand how societies operate and that a better understanding of organs, tissues, and molecules will teach us how organisms, and hence societies, operate. For this reason they explain the existence of crime on the basis of a "criminal personality" and believe that criminals behave as they do because they have diseased brains, too much or too little of certain hormones or other critical substances, or defective genes. So, reductionism is a bottom-up, hierarchical theory.

The converse is sometimes called *holism*. Some people base it on a similar analysis that accepts hierarchies of levels, but they assign superior explanatory power to the "higher" levels, the organism as a whole or even the organism in its surroundings. At present, this is a less popular system of explanation among scientists but one that carries considerable weight among practitioners of "alternative" methods of healing, such as massage and acupuncture, and among some feminists and environmentalists. They see reductionist ways of conceptualizing how organisms function and live in nature as a threat to people and our environment because these treat specific points of interest as though they could be isolated from the context in which they are embedded.

Biodeterminism is a form of reductionism in that it explains the behavior of individuals and characteristics of societies in terms of biology. Feminists know it best in the form of Freud's notorious statement that "biology is destiny." In previous chapters, we have encountered biodeterminist explanations for the obvious differences in women's and men's access to social, economic, and political power. Among them were Darwin's descriptions of the effects of sexual selection on the social behaviors of females and males. Biodeterminism also inspired nineteenth-century comparisons between the sizes of men's and women's brains and between brains of men of different races, which unequivocally "proved" the superiority of Caucasian men over men of other races and over all women.

A good deal of present-day research into the causes of social and behavioral differences between women and men relies on reductionist explanations that draw on hypothetical differences in hormone levels of female and male fetuses or on hypothetical genes that favor spatial skills, mathematical ability, competitiveness, and aggression in men and domesticity and nurturance in women. But the most pervasive and comprehensive of present-day biodeterminist theories is sociobiology, which has as its project "the systematic study of the *biological basis* of all social behavior" (Wilson, 1975, p. 4; my italics).

Sociobiological Models of Human Nature

Sociobiologists claim that it is possible to identify the fundamental elements of human nature by the fact that they characterize all people, whatever their cultural or historical differences, and selected animals as well. Once supposedly universal traits have been identified—for example, male aggression and female nurturance—sociobiologists argue that their very universality is evidence that they are adaptive.

The term *adaptive* has a special meaning in this context. It implies that the traits in question are inherited unchanged by successive generations and that individuals who exhibit them leave more descendants than other individuals do. In this way, the genes for more adaptive traits come to outnumber the genes for less adaptive ones until the more adaptive traits become universal. Sociobiologists argue that we try to do the things that help spread our genes about and behaviors that let us do that most effectively become universal. Prominent among such traits among males, they say, are behaviors that lead them to inseminate as many females as possible, hence promiscuity; among females, behaviors that optimize their ability to spot, and attach themselves to, genetically well-endowed males and to take good care of their offspring, hence fidelity and nurturance.

Sociobiologists believe that women's disproportionate contributions to the care of their children and homes are biologically programmed because women have a greater biological "investment" in children than men have. For this belief they offer the following rationale: An organism's biological fitness, in the Darwinian sense, depends on producing the greatest possible number of offspring who themselves survive long enough to reproduce. The number of offspring who reproduce determines the frequency with which an individual's genes will be represented in successive generations. Following this logic a step further, sociobiologists argue that women and men must adopt basically different strategies to maximize opportunities to spread genes into future generations.

As we have seen in the preceding chapter, the calculus goes as follows: Eggs are larger than sperm, and women can produce many fewer of them in a lifetime than men can sperm. Hence women are said to be the "scarce resource." Also, each egg that develops into a child represents a much larger fraction of the total number of children a woman can produce, hence of her total "investment" and "reproductive fitness" than a sperm that becomes a child does of a man's "investment" and "fitness." Furthermore, women "invest" nine months of pregnancy in each child, whereas men's procreative efforts are complete once they ejaculate. For these reasons, women must be more careful than men to acquire genetically well-endowed sex partners, who will also be good providers and will help women make their few investments (read, children) mature. Thus, from the seemingly innocent asymmetries between eggs and sperm flow such major social consequences as female fidelity, male promiscuity, women's disproportional contribution to the care of children, and the unequal distribution of labor by sex.

Such explanations ignore the fact that human societies do not operate with a few superstuds and that stronger and more powerful men, in general, do not have more children than weaker ones do. In theory, men could indeed have many more children than women can; but in most societies roughly the same number of women and men engage in producing children, although not in taking care of them. These kinds of biodeterminist theories are useful to people who have a stake in maintaining present inequalities, but they cannot help us explain social and economic realities.

There is no reason to believe that females expend more "energy" (whatever that means) on the biological components of procreation than males do. Females indeed produce fewer eggs than males do sperm, and, among mammals, females gestate embryos; but it is not obvious how to translate these facts into expenditures of energy. Is it reasonable to count only the energy required to produce the few sperm that actually fertilize eggs or should one not count the total energy males expend in producing and ejaculating semen (that is, sperm plus spermatic fluid) in a lifetime, however one might calculate that?

There are other puzzles. Does it make sense to describe the growth of a fetus inside its mother's womb as an investment of her energy? True, the metabolism of a mammalian embryo is part of a pregnant female's metabolic functions. As she eats, breathes, and metabolizes, some of the food and oxygen she takes in get used by the growing embryo. But why does that represent an investment of her energy? An embryo that grows in an undernourished woman is a drain because it will use her body for its growth. But well-nourished, healthy women often feel energized by their pregnancies and have no trouble living normal, active lives. They have even been known to compete in Olympic events.

To speak of the fetus as a drain on a pregnant woman's energies is reminiscent of the way physicians in the nineteenth century spoke of menstruation as requiring "energy" when they argued that girls who taxed their brains by becoming educated would not be able to have children when they grew up.

Sociobiological explanations that posit differences in the "energy" women and men "invest" in procreation to account for the fact that most men take less responsibility for the care of their children than most women do have a similar, superficial ring of scientific plausibility. But there is no way to specify the variables, much less to do the necessary calculations, to turn such waves of the hand into scientific statements.

Richard Dawkins (1976) takes sociobiological reductionism to its extreme by asserting that an organism is merely a gene's way of making more genes. He claims that everything organisms do they do out of self-interest because organisms are only living manifestations of "selfish genes" engaged in the process of replicating themselves. As we saw in chapter 6, one of the problems with this kind of formulation is that genes do not replicate themselves. Nor do eggs or sperm. Even many organisms do not reproduce themselves—at least not organisms that procreate sexually, such as people and most other animals sociobiologists discuss. In sexual procreation, individuals with different genetic makeups produce other individuals who are genetically different from both their parents and from each other. These differences have made it difficult for biologists to know how to analyze the ways in which even simple Mendelian traits, which involve differences in one gene, become established in a population. When it comes to the ill-defined behaviors that sociobiologists label selfishness, aggression, or nurturance, there is no rigorous way to determine that they are transmitted biologically or how they become established in populations.

Human sociobiology allows far too much leeway for equating traits that are observed in different cultures and under different historical circumstances. By calling different behaviors by the same name, sociobiologists make it appear as though they represented universal traits, especially when they go on and attribute the "same" traits to animals. In this way, everything from sharp business practices and warfare, to the roughhousing of toddlers and young animals, to interactions scientists observe among animals in the field, in zoos, or in crowded laboratory cages becomes "aggression." The term *rape*, which in ordinary speech refers to the violent, sexualized assertion of power men impose on unconsenting women and occasionally on other men, is used by sociobiologists merely to mean a way for males to spread their genes around. As a result, sociobiologists have described what they choose to call rape among birds, fishes, insects, and even plants (Barash, 1979). Contexts and cultural meanings are erased, and all that is left are reified traits, rendered universal because a variety of behaviors are called by the same name.

Because people are biological organisms, there are similarities between some of the ways animals and people behave. But the variety of animal behaviors is so great that one can ascribe evolutionary significance to any human behavior if the criterion is merely that some animals behave that way. That brings me to a crucial problem for scientific attempts to interpret in evolutionary terms relationships among the behaviors of different kinds of animals or of animals and people.

Ordinarily, when biologists try to understand whether anatomical similarities between different species are of evolutionary and hence genetic significance, they distinguish between two kinds of resemblances: analogies and homologies. Analogous traits are examples of what biologists call convergent evolution—different evolutionary pathways that provide similar solutions to similar biological or environmental problems. Examples of analogous structures are the wings of bats, birds, and insects, or the eyes of mammals, insects, and octopus or squid. Although they look similar and serve the same function—in our example, flight or vision—they have evolved independently and are not related genetically. Traits that share a common evolutionary and genetic basis are called homologous. Their common descent often is not obvious on superficial inspection. It must be deduced by careful, systematic examination of the paleontological record. Examples of homologous structures are the scales of reptiles and the feathers of birds. Although they look different, they serve similar functions (covering, insulation); but, most important, they can be shown to have a common ancestry.

To establish lines of biological inheritance, analogies are irrelevant. One must look for homologies, which usually requires culling the fossil record. But behavior leaves no fossils. There are only observations of how contemporary animals (including people) act and interpretations of what their behavior signifies. This kind of inspection offers much leeway for postulating connections and lines of descent on which to base hypotheses about why particular groups of people and animals behave in similar ways, but no solid information.

Interactive, Dialectical, and Complementary Models of Nature

I want to turn to a fundamental criticism of the hierarchy-of-levels analysis of nature, whether it ends up being interpreted in a reductionist or holistic manner—that is, bottom-up or top-down. To get away from reductionism and futile arguments about whether nature or nurture is more significant in shaping behavior, a number of scientists have stressed that, obviously, both genetics and environment are important. But more than that, genetic and environmental effects cannot be separated. According to the simplest model, the effects are additive. On the basis of this kind of model psychologists like Arthur Jensen (1969) and Richard Herrnstein (1971) have argued that 80 percent of intelligence is inherited, 20 percent is due to environment.

Other scientists point out that this formulation is overly simplistic, that nature and nurture interact in ways that cannot be quantified that easily because they are not additive but act simultaneously and affect each other. For example, Richard Lewontin (1974) has pointed out that we can assess the individual contributions of genetic and environmental factors that act jointly only under strictly controlled conditions that permit the experimenter to change one variable at a time. On the basis of such experiments scientists can construct graphs, called norms of reaction, that describe how specific changes in each variable affect the phenomenon under observation (such as the growth of a plant in various types of soil or in the same soil but under various conditions of moisture, temperature, or cultivation). But such norms of reaction do not permit one to predict the response of different varieties of the same organism even when grown under identical experimental conditions, or the reactions of a single variety under conditions that one has not measured. In other words, norms of reaction illustrate the complexity of the interrelationships but cannot be used to described the real world in which changes do not occur one variable at a time or in controlled or controllable ways.

Lewontin, Rose, and Kamin (1984), as well as Lynda Birke (1986), argue that this kind of interactive model, although less limited than the simple, additive ones, is still too static. Lewontin and his colleagues propose a dialectical model that acknowledges levels of organization, such as the ones I have enumerated, but makes none of them more fundamental than any other. None "causes" or "determines" another. Rather, all of them are related dialectically and mutually draw on, and

modify, the changes that may be produced at any particular
level. What is more, the properties at a particular level cannot
be deduced from properties observed at the other levels because all the levels are related dialectically. One cannot predict
the physics and chemistry of water from the properties of hydrogen and oxygen atoms. Nor can one predict the structures
and functions of proteins from the properties of the amino
acids of which they are composed, and even less from the properties of the atoms that make up the amino acids. Such predictions are impossible not because we do not know enough
about atoms or amino acids but because new properties emerge
when atoms or amino acids come together in different combinations, and these properties must be discovered empirically.
The same goes for the relationships between organisms and
their genes, or between societies and the individuals who live
in them.

I like to call this *transformationism,* an awkward term but one
that expresses the fact that biological and environmental factors can change an organism so that it responds differently to
other, concurrent or subsequent, biological or environmental
changes than it would have done otherwise. Simultaneously the
organism transforms the environment, which, of course, includes other organisms.

We can visualize this kind of dialectical interaction, or transformation, by thinking about the interplay between biological
and cultural factors that affect the ways boys and girls grow up
in our society. If a society puts half its children in dresses and
skirts but warns them not to move in ways that reveal their
underpants, while putting the other half in jeans and overalls
and encouraging them to climb trees and play ball and other
active outdoor games; if later, during adolescence, the half that
has worn trousers is exhorted to "eat like a growing boy," while
the half in skirts is warned to watch its weight and not get fat; if
the half in jeans trots around in sneakers or boots, while the
half in skirts totters about on spike heels, then these two groups
of people will be biologically as well as socially different. Their
muscles will be different, as will their reflexes, posture, arms, legs
and feet, hand-eye coordination, spatial perception, and so on.
They will also be biologically different if, as adults, they spend
eight hours a day sitting in front of a visual display terminal

or work on a construction job or in a mine. I am not saying that one is more healthful than the other, only that they will have different biological effects. There is no way to sort out the biological and social components that produce these differences, therefore no way to sort nature from nurture, when we confront sex differences or other group differences in societies in which people, as groups, do not have equal access to resources and power and hence live in different environments.

Recently, some of us have begun to use yet another model to look at the different levels of organization, one that draws on Bohr's principle of complementarity. Bohr proposed complementarity as a way to think about the fact that light and other electromagnetic radiation can be described as bursts of particles (quanta) and also as waves spreading out from a point source. Classical physicists argued over which they really were. Bohr and the other quantum theorists asserted that they were both, and by complementarity Bohr meant that they were both at all times, not sometimes one, sometimes the other. Which description is appropriate depends on the instruments an observer uses to examine the radiation: observed with a photoelectric cell, light is a random succession of packets of energy (quanta); observed with a diffraction grating or a prism, light is composed of waves.

Complementarity provides a fruitful model for integrating the different levels of organization we can use to describe living organisms. The phenomena we observe at the subatomic, atomic, molecular, cellular, organismic, or societal levels are all taking place simultaneously and constitute a single reality. It is an outcome of Western cultural history and of the history of professionalization that we have developed separate academic disciplines to describe these levels as though they were different phenomena. In fact, the only reason we think in terms of these levels is that we have developed specialties that draw distinctions between them. But the distinctions are not part of nature. Physicists and theoretical chemists, who calculate energy levels in atoms and molecules, do not have access to more fundamental truths than have molecular biologists, who study the structure and sequence of genes on the chromosomes. Nor are the descriptions molecular biologists provide more fundamental than those of biologists who study cells or organisms. Biologists do not probe deeper realities than anthropologists and historians, just different ones. The fact that academic professionals value the explanatory power of these disciplines differently tells us more about the history and sociology of professionalization and about the alliances different academic disciplines have been able to forge with economic and political power than about nature.

Problems with the Concept of Human Nature

It is questionable whether the concept of human nature means anything. People's "nature" can be described only by looking at the things they do. To try to abstract, or reify, a human essence from the ways in which different groups of people have grappled with issues of survival in the range of geographical, ecological, and demographic settings that our species has populated is a dubious enterprise because what people interpret as "natural" depends on their experience and viewpoint and is not likely to be agreed on by individuals with different backgrounds or interests. Margaret Mead (1949) pointed out years ago that societies with different, even opposite, sexual divisions of labor all believe that women's and men's tasks—whatever they are—follow from differences between women's and men's nature.

Prime among the traits sociobiologists believe to be inherent in our nature, as I pointed out before, is selfishness because that is what supposedly gets us to perpetuate our genes in the first place. A variant of selfishness is altruism—the kind that benefits the altruist in the end (something like "I'll scratch your back if you'll scratch my children's"). Then there are territoriality and a tendency toward establishing dominance hierarchies. These traits entered descriptions of animal behavior around the beginning of World War I, when a pecking order was first noted among barnyard chickens, a not very "natural" population. There are also the supposedly sex-differentiated characteristics of male aggressiveness and competitiveness and female coyness and nurturance that follow from the assumed asymmetry in our reproductive interests that I questioned previously (the fact that males produce a lot of sperm in a lifetime, whereas females produce relatively fewer, larger, and hence "more expensive" eggs). E. O. Wilson (1978) includes in "human nature" the various behaviors that make sexual relationships between women and men emotionally satisfying, such as fondling and kissing; religious and spiritual aspirations that generate the need to believe in something beyond oneself; and the incest taboo. (Apparently Wilson is unaware of the feebleness of this taboo, considering the widespread occurrence of sexual abuse in families, especially of girls by their fathers, brothers, uncles, and even grandfathers.) He acknowledges cultural influences but insists that biology contributes a "stubborn kernel" that "cannot be forced without cost" (p. 147).

Because sociobiologists posit that "stubborn kernel" of biological traits, honed over eons of evolution, their human-nature theories are conservative. They portray as natural the competitive and hierarchical capitalist societies in which men dominate women and a small, privileged group of men dominates everyone else. But competition and dominance hierarchies are not characteristic of all human societies, and there is no reason to believe that biology determines the ways different societies are constructed. On the contrary, people can undergo substantial physical as well as psychological changes as a result of major political and economic transformations in their societies. For example, rationing and other social policies the British government enacted during World War II resulted in a generation of people from working-class backgrounds who were healthier and looked significantly different from their parents. People who participate in major political or personal changes that drastically alter the ways they live often experience simultaneous changes in the ways their bodies function—changes in their ability to work and concentrate, in sleep and eating patterns, muscle mass, shape and strength, body weight, skin color and texture, and many other physical characteristics. It is not that changes in our way of life cause our biology to change. All the changes are interconnected: We change. Women who have participated in the women's liberation movement are well aware that such changes can occur in our "natures" as well as in our lives.

9

Rethinking Women's Biology

WOMEN'S biology is a social construct and a political concept, not a scientific one, and I mean that in at least three ways. The first can be summed up in Simone de Beauvoir's (1953) dictum "One isn't born a woman, one becomes a woman." This does not mean that the environment shapes us, but that the concept, woman (or man), is a socially constructed one that little girls (or boys) try to fit as we grow up. Some of us are better at it than others, but we all try, and our efforts have biological as well as social consequences (a false dichotomy because our biological and social attributes are related dialectically). As we discussed in the previous chapter, how active we are, what clothes we wear, what games we play, what we eat and how much, what kinds of schools we go to, what work we do, all affect our biology as well as our social being in ways we cannot sort out. So, one isn't born a woman (or man), one becomes one.

The concept of women's biology is socially constructed, and political, in a second way because it is not simply women's description of our experience of our biology. We have seen that women's biology has been described by physicians and scientists who, for historical reasons, have been mostly economically privileged, university-educated men with strong personal and political interests in describing women in ways that make it appear "natural" for us to fulfill roles that are important for their well-being, personally and as a group. Self-serving descriptions of women's biology date back at least to Aristotle. But if we dismiss the early descriptions as ideological, so are the descriptions scientists have offered that characterize women as weak, overemotional, and at the mercy of our raging hormones, and that construct our entire being around the functions of our reproductive organs. No one has suggested that men are just walking testicles, but again and again women have been looked on as though they were walking ovaries and wombs.

We have seen that in the nineteenth century, when women tried to get access to higher education, scientists initially claimed we could not be educated because our brains are too small. When that claim became untenable, they granted that we could be educated the same as men but questioned whether we should be, whether it was good for us. They based their concerns on the claim that girls need to devote much energy to establishing the proper functioning of their ovaries and womb and that if they divert this energy to their brains by studying, their reproductive organs will shrivel, they will become sterile, and the race will die out.

This logic was steeped in race and class prejudice. The notion that women's reproductive organs need careful nurturing was used to justify excluding upper-class girls and young women from higher education but not to spare the working-class, poor, or black women who were laboring in the factories and homes of the upper class. If anything, these women were said to breed too much. In fact, their ability to have many children despite the fact that they worked so hard was taken as evidence that they were less highly evolved than upper-class women; for them breeding was "natural," as for animals.

Finally, and perhaps most importantly, our concept of ourselves is socially constructed and political because our society's interpretation of what is and is not normal and natural affects what we do. It therefore affects our biological structure and functioning because, as I have said before, what we do and how our bodies and minds function are connected dialectically. Thus norms are self-fulfilling prophecies that do not merely describe how we are but prescribe how we should be.

Body Build and Strength

Let us consider a few examples. We can begin with a few obvious ones, such as height, weight, and strength. Women and men are physically not very different. There are enormous overlaps between women and men for all traits that are not directly involved with procreation.

For example, there is about a two-foot spread in height among people in the United States, but a difference of only three to five inches between the average heights of women and men. When we say men are taller than women, what we really mean is that the abstraction *average (or mean) height* is a few inches greater for men than women. Overall, women and men are about the same height, with many women as tall as, or taller than, lots of men. The impression that women are shorter than men is enhanced by our social convention that when women and men pair off, it is considered preferable for the man to be taller than the woman. In some countries, such as Bali, differences in height and, indeed, overall body build are much smaller than in the United States (Lowe, 1982).

Clearly, height is affected by social factors, such as diet. In the early part of this century, English working-class men were significantly shorter, on average, than men from the upper class, and this difference in height was due to differences not just in the adequacy but in the composition of their diets—proportions of carbohydrates, proteins, fats, vitamins. In the United States we are familiar with a similar phenomenon when comparing the heights of immigrants and their U.S.-born children. We have tended to think that the U.S.-born children are taller than their immigrant parents because they get a better diet. But now that we are learning more about the health hazards of the typical U.S. diet, with its excessive fat and protein content, we should probably defer value judgments and just acknowledge that the diets are different.

Sex differences in height probably also arise from the differences in growth patterns between girls and boys. Until early adolescence, girls, on average, are taller than boys, but girls' growth rates tend to decrease after they begin to menstruate, whereas boys continue to grow throughout their teens. It is generally assumed that this difference is due to the fact that the increase in estrogen levels after the onset of menstruation tends to slow the growth of girls' long bones. But the age of onset of menstruation, hence of increased estrogen secretion, depends on a number of social factors, such as diet, exercise and stress (Frisch, 1988). For example, female swimming champions, who, because of their intense, early training, tend to begin to menstruate later than most girls, tend also to be taller than average.

We might therefore expect factors that delay the onset of menstruation to decrease the difference in average height between women and men, those that hasten the onset of menstruation to increase it.

It is probably not that simple because the factors that affect the onset of menstruation may also affect height in other ways. All I want to suggest is that height, in part, is a social variable and that differences in the average height of women and men vary with the social environment.

Weight clearly has considerable social components. Different societies have different standards of beauty for women, and many of these involve differences in desirable weight. Today we call the women in Rubens's paintings fat and consider Twiggy anorexic. In our society changes in style not just of clothing but of body shape are generated, at least in part, because entire industries depend on our not liking the way we look so that we will buy the products that promise to change it. To some extent this is true also for men: Padded shoulders are not that different from padded bras. But there is more pressure on women to look "right," and what is "right" changes frequently and sometimes quite drastically. At present, U.S. women are obsessed by concerns about their weight to the point where girls and young women deliberately eat less than they need for healthy growth and development.

Although we may inherit a tendency toward a particular body shape, most women's weight can change considerably in response to our diets, levels of physical activity, and other patterns of living. These also affect physical fitness and strength. When women begin to exercise or engage in weight training and body building, we often notice surprisingly great changes in strength in response to even quite moderate training. Here again, what is striking is the variation among women (and among men).

People ask whether there are "natural" limits to women's strength and therefore "natural" differences in strength between women and men. In Europe and the United States women and men are far more similar in lower body strength than in the strength of our upper bodies. This fact is not surprising when we consider the different ways girls and boys are encouraged to move and play from early childhood on. We tend to use our legs much more similarly than our arms. Both girls and boys tend to run a lot, and hopscotch and skipping rope are considered girls' games. But when it comes to carrying loads, playing baseball, and wrestling and other contact sports, all of which strengthen the arms and upper body, girls are expected to participate much less than boys are. In general, male/female comparisons are made between physically more highly trained men and less trained women so that so-called sex differences at least in part reflect this difference in activity levels. More and less active men also differ in strength, and so do more and less active women.

If we compare the records of male and female marathon runners, we find that in 1963, when women were first permitted to run the Boston marathon, their record was 1 hour 20 minutes slower than the men's record. Twenty years later, Joan Benoit won in 2 hours, 22 minutes, and 43 seconds, a record that was only about fifteen minutes slower than the record of that year's male winner. And she ran the course in over an hour less time than the female winner in 1963 (Fausto-Sterling, 1985). The dramatic improvement women runners made in those twenty years clearly came with practice but no doubt also required changes in their expectations of what they could achieve. Men's records have improved by less than fifteen minutes during the entire time since modern marathon competitions began in 1908. Again the question: Are there "natural" limits and "natural" differences between women and men? Only time and opportunities to train and to participate in athletic events will tell. Note that in the 1988 Olympics, the woman who won the hundred-meter sprint took less than one second longer than the male winner, and he set a new world record. This feat is especially remarkable because women are said to compare with men much more favorably in long runs than in sprints.

Work

The stratification of the work force is often explained as though it reflected inherent biological differences between women and men. Women have been disqualified from construction and other relatively well-paid heavy labor because they are said to be too weak for it. But the most prestigious men's jobs and those that pay most, in general, do not require physical strength, while much of women's traditional, unpaid or underpaid work involves strenuous physical labor. Nurses must sometimes lift heavy, immobilized people, and housework frequently involves carrying and pushing heavy, awkward loads. In many cultures women are responsible for providing the firewood and water, which usually means carrying heavy loads for long distances, often with small children tied to their chests or backs. In the United States, where men are expected to carry the heaviest loads, most men have "bad backs," which is why occupational health advocates argue that loads that are considered too heavy for women should be rated too heavy for everyone.

At present, there is an overemphasis on the reproductive hazards of employment for women and an underemphasis on comparable hazards for men, to the detriment of women, men, and children. As we saw in chapter 2, women have been barred from some higher-paying jobs unless they could prove they were sterile, while men in those very jobs, and in others, continue to be exposed to preventable chemical and radiation hazards (Stellman and Henifin, 1982). Women, too, continue to be exposed to reproductive hazards in traditional women's work, such as nursing and housework, and as x-ray technicians, beauticians, and hairdressers.

In other words, biological differences between women and men are used to rationalize the stratification of the labor force by sex; they do not explain it. One can readily find women or men who qualify for every kind of paid work, except that of sperm donor and what has come to be called surrogate mother. If society instead stratifies the work force into women's and men's jobs, it does so for economic, social, and political reasons. Such stratification is not mandated by biology.

Menstruation

Let us leave these biosocial examples and look at menstruation, which most people would consider purely biological. A good way to begin is by asking, What is a normal woman's normal menstrual pattern? The standard answer is: a twenty-eight day cycle with five days of menstruation that begins at age twelve or thirteen and ends at about fifty. Yet that pattern does not reflect most women's actual experience. Until recently, we have had little information about the normal range of variation in age of onset, frequency, regularity, discomfort, and cessation of menstruation. Little information has been shared among women, and there has been almost no research on our routine experiences. Women have learned about menstruation mostly from their mothers or other female relatives or, if they had problems, from physicians—most of them men, who learn what they know from textbooks written by other men or from their "clinical experience," which means from women with problems.

In recent years, women's health activists and feminist medical and social scientists have finally begun to give us a sense of the variety of women's normal experiences of menstruation (Boston Women's Healthbook Collective, 1984; Martin, 1987). We are also beginning to learn about the experiences in other times and cultures. Rose Frisch (1988) has shown that, during the last century, the age of onset of menstruation has gone down and the age of cessation has gone up in both Europe and the United States. From this change and from studies of the menstrual patterns of athletes, she has concluded that nutrition and exercise strongly affect these parameters, probably by influencing the amounts of stored body fat. She suggests that women need to accumulate a threshold amount of fat in order to establish the hormonal cycles that regulate menstruation. As European and American diets have become richer in fats, girls reach the critical level earlier, and older women maintain it longer. Participation in vigorous sports affects menstrual patterns because athletes convert more of the food they eat into muscle (protein) and store less of it as fat.

Anthropologists observing the !Kung, a group of foragers living in the Kalahari desert in southern Africa, have noted that their menstrual and reproductive histories are quite different from what we in the West think of as "normal" (Howell, 1979; Konner and Worthman, 1980; Shostak, 1981). !Kung women and men collect their food, and as is true in most foraging societies, women provide most of it which involves a good deal of walking and carrying. The !Kung diet is plentiful and nutritionally adequate but very different from ours because it is relatively high in complex carbohydrates and plant proteins but low in animal proteins and fats.

Presumably because of their high activity levels and their diet young !Kung women do not begin to menstruate until they are about eighteen years old, by which time they already tend to be heterosexually active. Like girls in the West, they tend not to ovulate during their first few cycles. They therefore experience their first pregnancy when they are about nineteen and have a first child at perhaps twenty. They nurse that child for two or three years but quite differently from the way many of us do. They suckle their babies as often as the infant wants to nurse, which can be several times an hour, albeit briefly. Melvin Konner and Carol Worthman (1980) have postulated that because of the frequent nursing, !Kung women tend not to menstruate or ovulate for almost the entire time they suckle their children. This experience contrasts with that of Western women, who tend to resume menstruation within a year after a birth, even when they nurse their children for several years, because they let them suck much less frequently and tend to supplement their diet with other foods. Thus, Western babies suckle less intensely and frequently than !Kung babies do.

!Kung women tend to wean a child sometime during its third year. By this time they may be pregnant, without having resumed menstruation, or they may menstruate and ovulate a few times before they become pregnant again. This pattern repeats until the women reach menopause, which they tend to do in their late thirties or early forties.

The menstrual and reproductive experience of !Kung women therefore is entirely different from what we take to be "normal." They have a shorter reproductive span, during which they tend to bear no more than four or five children, quite without contraception, and they experience few menstruations. But when the !Kung move into towns and live more as we do, their menstrual and reproductive patterns change to the ones we are used to seeing. So the difference between their experience and ours is not genetic.

Clearly it is meaningless to specify norms for even such a normal, biological function as menstruation without considering how women live. For !Kung women it is normal to menstruate rarely and have few children without using birth control. For us it is normal to menstruate every twenty-eight or so days and to get pregnant within a year after a birth, if we are heterosexually active without using birth control. Thus, even such biological events as menstruation and fecundity are strongly influenced by sociocultural factors.

Women's Biology in Context

How had we best take these influences into consideration? Clearly we need to think about women's biology in its social context and consider how it interacts with culture. We need to get information directly from women and not rely on so-called experts, who are often male and whose knowledge tends to be based on the experience of "patients"—that is, of women with problems. Only when women have the opportunity to share experiences and when scientists collect the experiences of women of different ages and from different classes, races, and cultural groups can we get a sense of the texture and variety of women's biology (Martin, 1987).

We need to pay attention to the obvious contradictions between stereotypic descriptions of women's biology and the realities of women's lives. For example, women's reputed "maternal instinct" needs to be looked at in light of some women's desperate efforts to avoid having children, while society persuades or forces them to have children against their wills. Similarly, descriptions of women's frailty, passivity, and weakness need to be juxtaposed with the reality of women as providers and workers who in most societies, including our own, tend to work harder and for longer hours than most men.

Women's work histories are often obscured by the fact that work has been defined so that it excludes much of their daily work load. Indeed, whereas most of what men do is called work, much of what women do has been interpreted as the natural manifestation of our biology. How often do we hear people say, "My mother didn't work when I was growing up"? If she didn't work, how did we manage to grow up? Even women usually

refer to what we do as work only when we get paid for it, implying that what we do at home and in our neighborhoods and com-
munities is not work. This misrepresentation of work sets up
the vicious circle whereby women are thought to be less good
workers in the workplace when we have family and community
obligations and less good housewives and mothers when we
work outside the home.

No question, we are biological organisms like other animals,
and women and men have different procreative structures and
functions. But to try to find the biological basis of our social
roles or to sort people by sex when it comes to strength, ability
to do math, or other intellectual or social attributes is a political
exercise, not a scientific one.

The Meaning of Difference

That said, I want to stress that we need have no ideological investment in whether women and men exhibit biological differences, aside from the obvious ones involved with procreation.
I have argued that we cannot know whether such biological differences exist because biology and society (or environment) are
interdependent and cannot be sorted out. And in any gender-
dichotomized society, the fact that we are born biologically female or male means that our environments will be different:
We will live different lives. Because our biology and how we live
are dialectically related and build on one another, we cannot
vary gender and hold the environment constant. Therefore,
the scientific methodology of sex-differences research is intrinsically flawed if scientists try to use it to sort effects of biology
and society. Scientists can catalog similarities and differences
between women and men but cannot establish their causes.

There are other problems with research on differences. One
is that it is in the nature of scientific research that if we are interested in differences, we will go on looking until we find
them. And if we do not find any, we will assume that our instruments were wrong or that we looked in the wrong place or at
the wrong things. Another problem is that most characteristics
vary continuously in the population rather than placing us into

neat groups. To compare groups, however defined, we must use such concepts as the "average," "mean," or "median" in order to characterize each group by a single number. Yet these constructed, or reified, numbers obscure the diversity that exists within the groups (say, among women and among men) as well as the overlaps between them. That is why statisticians have invented the concept of the standard deviation from the mean to reflect the spread of the actual numbers around the reified average. This problem is obvious when we think about research into differences between blacks and whites. Just to do it, we have to agree on social definitions of who will count as black and who as white because after several centuries of mixing, the biological characteristic, skin color, varies continuously. Research comparing blacks and whites must first generate the group differences it pretends to catalog or analyze.

Differences, be they biological or psychological, become scientifically interesting only when they parallel differences in power. We do not frame scientific questions about differences between tall people and short people, although folk wisdom suggests there may be some. Nor do we, in this society, pursue differences between blue-eyed, blond people and dark-haired, dark-eyed ones. Yet the latter were scientifically interesting differences under the Nazis.

Sex differences are interesting in sexist societies that value one group more highly than the other. Because the overlaps are so large for all the characteristics that are not directly involved with procreation, it is easy to find women and men to perform any task we value. The existence of average sex differences is irrelevant to the way we organize society. To achieve an egalitarian division of labor requires political will and action, not changes in our biology. There is enough variability among us to let us construct a society in which people of both sexes contribute to whatever activities are considered socially useful and are rewarded according to their talents and abilities.

10

The Social Construction of Sexuality

THERE is no "natural" human sexuality. This is not to say that our sexual feelings are "unnatural" but that whatever feelings and activities our society interprets as sexual are channeled from birth into socially acceptable forms of expression.

Western thinking about sexuality is based on the Christian equation of sexuality with sin, which must be redeemed through making babies. To fulfill the Christian mandate, sexuality must be intended for procreation, and thus all forms of sexual expression and enjoyment other than heterosexuality are invalidated. Actually, for most Christians nowadays just plain heterosexuality will do, irrespective of whether it is intended to generate offspring.

These ideas about sexuality set up a major contradiction in what we tell children about sex and procreation. We teach them that sex and sexuality are about becoming mommies and daddies and warn them not to explore sex by themselves or with playmates of either sex until they are old enough to have babies. Then, when they reach adolescence and the entire culture pressures them into heterosexual activity, whether they themselves feel ready for it or not, the more "enlightened" among us tell them how to be sexually (meaning heterosexually) active without having babies. Surprise: It doesn't work very well. Teenagers do not act "responsibly"—teenage pregnancies and abortions are on the rise and teenage fathers do not acknowledge and support their partners and babies. Somewhere we forget that we have been telling lies. Sexuality and procreation are not linked in societies like ours. On the contrary, we expect youngsters to be heterosexually active from their teens on but to put off having children until they are economically independent and married, and even then to have only two or, at most, three children.

Other contradictions: This society, on the whole, accepts Freud's assumption that children are sexual beings from birth and that society channels their polymorphously perverse childhood sexuality into the accepted forms. Yet we expect our children to be asexual. We raise girls and boys together more than is done in many societies while insisting that they must not explore their own or each other's sexual parts or feelings.

What if we acknowledged the separation of sexuality from procreation and encouraged our children to express themselves sexually if they were so inclined? What if we, further, encouraged them to explore their own bodies as well as those of friends of the same and the other sex when they felt like it? They might then be able to feel at home with their sexuality, have some sense of their own and other people's sexual needs, and know how to talk about sexuality and procreation with their friends and sexual partners before their ability to procreate becomes an issue for them. In this age of AIDS and other serious sexually transmitted infections, such a course of action seems like essential preventive hygiene. Without the embarrassment of unexplored and unacknowledged sexual needs, contraceptive needs would be much easier to confront when they arise. So, of course, would same-sex love relationships.

Such a more open and accepting approach to sexuality would make life easier for children and adolescents of either sex, but it would be especially advantageous for girls (Jackson, 1982). When a boy discovers his penis as an organ of pleasure, it is the same organ he is taught about as his organ of procreation. A girl exploring her pleasurable sensations finds her clitoris, but when she is taught about making babies, she hears about the functions of the vagina in sex and birthing. Usually, the clitoris goes unmentioned, and she doesn't even learn its name until much later. Therefore for boys there is an obvious link between procreation and their own pleasurable, erotic explorations; for most girls, there isn't.

Individual Sexual Scripts

Each of us writes our own sexual script out of the range of our experiences. None of this script is inborn or biologically given. We construct it out of our diverse life situations, limited by what we are taught or what we can imagine to be permissible and correct. There is no unique female sexual experience, no male sexual experience, no unique heterosexual, lesbian, or gay male experience. We take the experiences of different people and sort and lump them according to socially significant categories. When I hear generalizations about *the* sexual experience of some particular group, exceptions immediately come to mind. Except that I refuse to call them exceptions: They are part of the range of our sexual experiences. Of course, the similar circumstances in which members of a particular group find themselves will give rise to group similarities. But we tend to exaggerate them when we go looking for similarities within groups or differences between them.

This exaggeration is easy to see when we look at the dichotomy between "the heterosexual" and "the homosexual." The concept of "the homosexual", along with many other human typologies, originated toward the end of the nineteenth century (d'Emilio and Freedman, 1988; Weeks, 1977). Certain kinds of behavior stopped being attributed to particular persons and came to define them. A person who had sexual relations with someone of the same sex became a certain kind of person, a "homosexual;" a person who had sexual relations with people of the other sex, a different kind, a "heterosexual."

This way of categorizing people obscured the hitherto accepted fact that many people do not have sexual relations exclusively with persons of one or the other sex. (None of us has sex with a kind of person; we have sex with a person.) This categorization created the stereotypes that were popularized by the sex reformers, such as Havelock Ellis and Edward Carpenter, who biologized the "difference." "The homosexual" became a person who is different by nature and therefore should not be made responsible for his or her so-called deviance. This definition served the purpose of the reformers (although the laws have been slow to change), but it turned same-sex love into a medical problem to be treated by doctors rather than punished by judges—an improvement, perhaps, but not acceptance or liberation.

Theories of Sexual Development

Freud was unusual for his time (and still is, to some extent, for ours) by insisting that sexual development is not innate and automatic. He considered it scientifically as valid to ask how people come to love individuals of the other sex as of their own. Nonetheless, he then plotted a course of "normal" development which involved his newly invented Oedipus complex, castration anxiety, and penis envy, to explain how men come to form affective attachments to women and women to men. Thus, loving people of one's own sex continued to be seen as pathological.

Feminist revisions of Freud by Nancy Chodorow (1978) and Dorothy Dinnerstein (1976) delineate affective development by putting the main emphasis on the maturing child's relationship to the mother rather than to the father, as Freud had done. Because a child's first loving relationship usually is with the mother or some other woman, for girls this relationship is with a person of the same sex, whereas for boys, with a person of the other sex. Therefore their analysis, like Freud's, posits a crucial difference between the ways girls and boys develop sexual identities and erotic relationships with members of the other sex.

Freud delineated a course that was clearer and more direct for boys and fuzzier and more problematical for girls. Chodorow and Dinnerstein suggest that male psychosexual development is the more problematical. They argue that because all children initially identify with their primary caretaker, who is usually a woman, girls can maintain their initial gender identification. Because boys must grow up to be men, they have to become ostentatiously unlike the person who cares for them and who was their first love. Yet boys, like girls, usually are not nearly so familiar with a man as they are with their female caretaker. This necessity to differentiate themselves in kind from the person they know and love best engenders a fragility in the male ego that women need not cope with. It is surprising that neither Chodorow nor Dinnerstein addressed the question of why, in that case, so many women later form affective ties with men rather than transferring their primary bond from the mother, or other female caretaker, to other women. Their model readily lends itself to suggesting that, from a developmental point of view, for both men and women love for women is less problematical than love for men. But they did not explore these implications.

Toward a Nondeterministic Model of Sexuality

I do not want to imply that they are worth pursuing in that form. I am no more comfortable with models that posit a psychological determinism, in which what happens in early childhood is all-important, than I am with biodeterminism. I find Chodorow's and Dinnerstein's analyses more interesting than Freud's but no more convincing. With sexuality, as with other aspects of our lives, I prefer to stress human diversity, flexibility, and our ability to change. Alfred Kinsey and his collaborators (1948, 1953) showed a long time ago that most people can love people of either sex and that our choices often change over time and with social circumstances.

Some gay men and lesbians feel that they were born "different" and have always been homosexual. They recall feeling strongly attracted to members of their own sex when they were children and adolescents. But many women who live with men and think of themselves as heterosexual also had strong affective and erotic ties to girls and women while they were growing up. If they were now in loving relationships with women, they might look back on their earlier loves as proof that they were always lesbians. But if they are now involved with men, they may be tempted to devalue their former feelings as "puppy love" or "crushes."

Even within the preferred sex, most of us feel a greater affinity for certain "types" than for others. Not any man or woman will do. No one has seriously suggested that something in our innate makeup makes us light up in the presence of only

certain women or men. We would think it absurd to look to hormone levels or any other simplistic biological cause for our
preference for a specific "type" within a sex. In fact, scientists
rarely bother to ask what in our psychosocial experience shapes
these kinds of tastes and preferences. We assume it must have
something to do with our relationship to our parents or with
other experiences, but we do not probe deeply unless people
prefer the "wrong" sex. Then, suddenly, scientists begin to look
for specific causes.

Because of our recent history and political experiences, feminists tend to reject simplistic, causal models of how our sexuality develops. Many women who have thought of themselves as
heterosexual for much of their life and who have been married
and have had children have fallen in love with a woman (or
women) when they have had the opportunity to rethink, refeel,
and restructure their lives.

The society in which we live channels, guides, and limits our
imagination in sexual as well as other matters. Why some of us
give ourselves permission to love people of our own sex whereas
others cannot even imagine doing so is an interesting question.
But I do not think it will be answered by measuring our hormone levels or by trying to unearth our earliest affectional ties.
As women begin to speak freely about our sexual experiences,
we are getting a varied range of information with which we can
reexamine, reevaluate, and change ourselves. Lately, increasing
numbers of women have begun to acknowledge their "bisexuality"—the fact that they can love women and men in succession or simultaneously. People fall in love with individuals, not
with a sex. Gender need not be a significant factor in our choice,
although for some of us it may be.

11

Constructing Sex Difference

We cannot understand literature and its study apart from the fact that we as authors, scholars, and critics are gendered, nor can we understand ourselves apart from the cultural organization of gender that produces us.

There is no neutral position from which we view subjects of study. Our contexts and theirs create what we see of them.
—Ellen Messer-Davidow,
"The Philosophical Bases of Feminist Literary Criticism"
(1987)

I want to let these two statements draw me into an exploration of the ways our society goes about constructing not only a gendered literature but sex and gender as such.

John Money and Anke Ehrhardt (1972) have popularized the distinction between sex and gender in their discussions of how children born with ambiguous genitalia develop a gender identity. (Nowadays such children have their gender reassigned, provided their genitals can be brought into accord with their proposed gender.) As Barbara Fried (1979) has pointed out, that distinction replaces the single, male/female, dichotomy with two: male/female, descriptive of sex and said to be based in biology, and masculine/feminine, descriptive of gender and said to be based in socialization. "And," writes Fried, "we are left to spend our time squabbling over whether each trait displayed by a man is more rightly attributed to his maleness (sex) or his masculinity (gender)." By mixing and matching these terms, society can use this dichotomy to enforce conformity. It can generate the "feminine man" and the "masculine woman," and so label as deviants women and men who behave "inappropriately" when there is no justification for calling their maleness or femaleness into question.

The distinction between sex and gender and these ways of using it are based on the implicit, but false, assumption that the effects of biology and society are discrete and separable, at least in theory if not in practice. As we have seen in the preceding chapters, these effects are in fact inseparable. Every organism constantly transforms its environment while being transformed by it, and, in the case of people, the society in which we live is a major component of our environment.

In our society sex and gender are primary areas of social molding. Much as I rebel against the ever-present dichotomy between female and male, I cannot meet someone who has just had a child without asking its sex. Our language forces me to because I cannot accord that child personhood and stop calling it "it" unless I know whether "it" is a "she" or a "he." And once I am told which it is, that information engenders (note the word) a host of meanings for me and the parents.

Margaret Mead (1949) taught us long ago that sex roles are malleable. What is sauce for the goose in one society may be sauce for the gander in a different one. But Mead also pointed out that whatever the specific roles prescribe—whether women or men are the ones to carry the heavier loads, whether men or women are supposed to express a wider range of emotions—people in that society believe these roles to be both natural and inevitable. In our society that translates into the belief that they are based in biology.

Clearly there are inborn sex differences: Most babies are born either with ovaries and a uterus, vagina, and clitoris or with testicles and a penis. An occasional child is born with anatomical discordances or ambiguities, but, to make things simple, let us confine our attention to the majority: babies who are born with concordant internal and external genitalia and are identified unambiguously as girls or boys. Each of us is also born with our specific, unique inheritance, with Aunt Mary's nose and Grandpa John's dimpled chin.

But, as we have seen before, what food we get and how much, what diseases we are exposed to, where we live, how physically active we are encouraged to be, how much time we spend with

other people, everything from our immediate family environment to whether the nation in which we live is at war, will affect
the ways our bodies develop, how strong we become, how alert,
how confident. And day by day, a toothache, a good night's
sleep, menstrual cramps, the way our muscles feel after a bout
of exercise or hard physical labor all affect the ways we interact
with the world around us.

Some of these effects may be transient, but other physical
characteristics (such as eye and skin color or genital anatomy) can have long-term effects. Despite the fact that our
environment can transform at least some physical, as well as
psychosocial, characteristics (which, of course, is another false
dichotomy), I do not want to imply that a baby is merely a passive recipient of manifold influences. Even relatively powerless
people, such as babies, act on their environment and transform
certain aspects of it consciously or unconsciously. So once again
the point is that we live in dynamic interaction with our environment. Sex differences are socially constructed because
being raised as a girl or a boy produces biological as well as social differences. Society defines the sex-appropriate behavior to
which each of us learns to conform, and our behavior affects
our bones, muscles, sense organs, nerves, brain, lungs, circula-
tion, everything. In this way society constructs us as biologically,
as well as socially, gendered people. It does not give us a vagina
or a penis, but it helps give us the muscles, gait, body language,
and nervous responses that we associate with people who are
born with one or the other.

Recent historians of sexuality have wondered why sex difference is of such central concern in our scheme of things (Foucault,
1978; Weeks, 1981). Although they do not have a definitive answer, it is clear that sex difference has not been emphasized in
all times and places and therefore need not be. In the article I
cited in the epigraph, Messer-Davidow suggests, I believe correctly, that the reason has to do with power relationships. Sex
and gender dichotomies lend themselves to setting up norms,
hence to inventing deviance. They act as instruments of social
control. This explanation, however, does not answer the question of why sex difference and sexuality are such important
arenas of social control in this society.

I assume that our preoccupation with matters of sex accounts for the fact that the most fundamental theory in biology—the theory of evolution by natural selection—is constructed around sex and procreation, and that psychoanalysis—still perhaps the most widely accepted psychological theory—is constructed around psychological interpretations of our experience of sex difference (in Freudian, anatomical reductionism, fear of castration and the Oedipus complex for boys, and penis envy for girls).

As we try to reconceptualize the role biology plays in shaping our gender identities, I find it informative to look at the changes Alice Miller (1984, 1986) has introduced into psychoanalytic thinking. Miller uses orthodox psychoanalytic methodology, complete with couch, dredging up of earliest memories, and reliving them by transference to the analytic therapist, but she denies the existence of innate drives, including sexual drives.

I disagree with Miller's assumption that our personality structure is determined in infancy and can be modified later only by means of psychoanalytic therapy. I believe that our psychological development (like our biology) is more flexible than that, that we have the potential for ongoing transformation. But, like Miller, I reject the notion that infants enter the world equipped with inborn sexual drives that must be resolved during childhood by means of psychosexual crises, such as castration anxiety and the Oedipus complex if they are boys, or penis envy if they are girls.

Miller asserts that children's sexuality, as well as their other so-called drives, develop because adults impose their own repressed, hence unconscious, needs on the infants they have in their care. This imposition is all the easier because infants depend on adults to introduce them into the culture. In a sexually repressed and neurotic society, such as ours, this dependency necessarily exploits the child. Children cannot reject the demands adults make on them when they need to have those adults love them as well as care for them. In other words, children are objects of manipulation by us adults (often loving and caring adults) who do not acknowledge our children as authentic human beings with legitimate needs of their own but instead use them to realize ours. And because our sexual needs tend to be the least conscious and acknowledged of our needs, no wonder we sexualize our relationships with our children.

Children are not passive bystanders in this drama. They have their own biological and psychological needs, which they engage in their relationships with the adults on whom they depend. But the basic asymmetry in power and capabilities between children and those of us who care for them makes children use the kinds of resources available to the weak. Feminists have described how women survive and succeed in male-dominated societies by means of "feminine wiles"—seductiveness, giggling, fawning, and so on. Similarly, children learn to live in the adult world by being "childish"—they are cute, whiney, throw tantrums.

I have sketched a relatively uncharted territory for us to explore if we want to understand the mechanisms that genderize and sexualize us. Perhaps the situation in our biosocial development is not as grim as the situation Miller describes for our psychosocial development. The adult world may have fewer preconceptions about children's gender-appropriate physique, body stance, and functioning than about gender-appropriate behavior. But even as I write these words, I am not sure. Infants are swaddled and their feet and other parts are bound figuratively if not in fact. The task I am urging on us is to become aware of the forced molding of our bodies as well as our minds, especially because, as feminists, most of us do not believe that our bodies and minds are separable. If we gain that awareness, we will not be in what Messer-Davidow describes as a "neutral position from which we [can] view subjects of study." That kind of awareness may instead put us in a position to understand and, if possible, overcome the ways genderization limits our critical insights and intuitions as well as our will to live in our bodies as fully as we can.

PART THREE

How Do We Use It?

Some of the most recent research in biology is being applied in the new procreative technologies. Prenatal diagnosis and screening are becoming part of routine prenatal care, and new technical interventions can help people who are unable to beget or conceive children or to carry pregnancies to term to have children who are biologically related to at least one of the future social parents. Some of the techniques can also help people without partners of the other sex to have biological children. Why the new, or surely increased, impulse to proceed into these areas? It would be a mistake to assume that this impulse is driven mainly by the availability of new biotechnologies. Technologies can open new possibilities, but we usually do not pursue them unless there are interest groups who want, and have the power to explore and implement, them. So what are the interests?

The Meaning of Procreative Choice

The ideology of procreative choice has become important in our society. In the course of the last hundred years, we in the middle and upper-middle class, who are used to being able to plan other aspects of our lives, have come to expect that we can plan our families. We decide how many children we want and when we are ready to have them. We practice contraception and, if it fails, abortion in order to implement these decisions.

Throughout we tend to assume that we will be able to have a child when we are ready. If we try and nothing happens, we feel not only distressed but wronged—and not by God or fate, as we might have in previous times, but by our bodies. In a time of artificial hearts and kidneys, we expect medical technology to be able to solve such problems. What is more, we do not want to have just any child. We want a healthy one because a child who needs more than the usual care or needs it for longer than usual will be hard to fit into our plans. Thus, prenatal diagnosis and the new technologies that let people have biological children have become part of planning our families.

Yet this kind of family planning is not a possibility for everyone. Owing largely to the work of women of color, the women's movement has begun to acknowledge that women are not a homogeneous group. Not only do we differ as individuals, but we fall into groups with common interests that may be different from those of other groups of women. We have finally realized that people who use the word *women* without qualifiers tend to focus their attention mainly on the young, white, fairly affluent women who became identified with the women's movement in the 1960s. These women have produced major changes; but if the women's movement is to continue to be a force for progressive change, it is essential that we acknowledge the differences in women's needs and interests.

It is not enough to address the issue of diversity without speaking explicitly about racism and class discrimination. And this need is nowhere more apparent than in our concerns about procreation, some of which I will discuss in the chapters that follow. The way feminists have framed the important issues in this area betrays the individualistic bias of the affluent, white, American upper-middle class. Our watchwords have been "reproductive freedom" and "choice," but we have not emphasized sufficiently that access to economic and social resources is essential to freedom of choice. In the United States poor women, who because of racism are disproportionately women of color, cannot count on having adequate housing, food, healthcare, jobs, and childcare. Yet all these necessities are basic to procreative freedom and choice. The most recent report of the Washington-based Children's Defense Fund (1989) documents the dismal economic situation in which large numbers of women and children in this country are living after many years of inflation, privatization, and cuts in government expenditures for human services.

Most people take the phrase *procreative choice* to mean the choice not to have children, surely an important concern for all women who have sex with men. But procreative choice also needs to mean the choice to have children in the confidence that we will be able to care for them. And that choice is not available to many poor women and usually is not even acknowledged as part of procreative choice.

In the late nineteenth and first half of the twentieth century, a scientific and social movement developed with the aim of denying procreative choice to certain kinds of people. This eugenics movement had as its aim furthering procreation among affluent, native-born, white Americans, while discouraging poor people, immigrants, and other "undesirables" from having children by persuasion and sometimes by forcible sterilization (Kevles, 1985; Ludmerer, 1972). The birth-control movement, under the leadership of feminists like Margaret Sanger, incorporated considerable portions of this ideology (Davis, 1981; Gordon, 1976).

In the chapters that follow I discuss some of the ramifications of prenatal diagnosis, fetal therapy, and in vitro fertilization. But we must acknowledge from the start that these technologies are not intended for all people who might feel they need them. They are expensive and often require that clients be sophisticated in the ways they relate to the medical system. Also, the techniques designed to enable people to have children are clearly intended for those who fit the stereotypic image of "the family" bread-winning dad, homemaking mom, and their children. They are not meant for poor people, hence not for the disporportionate number of people of color who are poor. They are not meant for lesbians or gay men. They certainly are not meant for people with disabilities. In fact, these people are high on the list of folks expected to use prenatal testing. They are supposed to avoid having children, and if they are so improvident as to want children, surely to do all they can to avoid having children who have disabilities.

Power, Risks, and Choice

In her novel *The Handmaid's Tale,* Margaret Atwood (1986)
illustrates how a society, without using any technology, can per-
vert women's procreative abilities so that they become altogether oppressive. But that does not mean that procreative
technologies are neutral and that only the ways they get used
determine whether they liberate or oppress. Complicated technologies cannot be developed or used without the help of experts. Ordinary people cannot get access to them or evaluate
their risks without the advice of trained professionals. In this
society a need for professionals means that the technology is expensive and introduces differences in power between those
who administer it and the users.

A look at the degree to which some of the procreative technologies differ in this regard will illustrate the point. Contraception is low-tech. We can use it with minimal medical
interference, and although we would like to have more choices
and more convenient and less health-threatening contraceptive
methods, our day-to-day use of the technology need not involve
us in power relationships with physicians. So contraception has
been liberating not only because we need it but because it is
largely in our hands. To perform abortions requires training,
although not professionals, and in this country we have so far
been able to keep the choice to have an abortion in our hands.
By contrast, in vitro fertilization is medicalized from the start. It
involves extensive preliminary tests, hormone treatments, and
usually several surgical procedures. And as we shall see in chapter 15, it has a low probability of success.

Prenatal diagnosis returns women to past times when people
looked on every pregnancy as a health risk. But as we shall see
in the next chapter, in the eighteenth and nineteenth centuries,
pregnant women were in fact risking disability and death. Now
that women no longer need have such fears, we are told to look
on every fetus as potentially disabled and in need of on-going
medical surveillance. But the reality is that only the rare fetus is
at risk for serious genetic or developmental problems as long as
pregnant women have access to adequate nutrition and the necessary economic and social supports, and can live in a relatively

healthful environment. Poverty, malnutrition, and urban decay place a fetus at far greater risk than do the inherited disabilities for which prenatal tests are being developed (Nsiah-Jefferson, 1988).

In the chapters that follow I shall look at the framework in which the new procreative technologies are being developed and used, not merely at their risks and benefits for individual users. I also want to consider how they affect women's experience of childbearing and how the larger society is coming to look on that experience.

Clearly the procreative technologies represent only one of the ways in which biological science and technology is coming to affect our lives. I concentrate on them not because I think they are more important than, say, applications of biotechnology to agriculture or warfare but because I believe that the current reshaping of the ideology and process of procreation offers poignant illustrations of how not to make decisions about what forms of science and technology to develop.

I want to make one point clear at the start. When I criticize the procreative technologies for their ideological content or their practical consequences, I do not mean to criticize, much less blame, the women who use them. I agree with Barbara Katz Rothman (1989) that the popular mythology about motherhood has changed drastically since the 1960s. The image of the selfless mother, who sacrifices her needs to those of her family, has been replaced by that of the selfish mother, who puts her own needs first and pits her interests against those of her fetuses and children. I take it for granted that childbearing is difficult in this society and many others. Women try to do it as best they can and get little support or credit.

The question for me is not whether women should use the available technologies if they think these will make their and their families' lives easier. I am concerned about the emotional costs of the technologies to the individual users as well as about their excessive financial costs. But more importantly, I question the process by which decisions are made about what technologies to develop. Such decisions should not be left to small groups of scientists who usually also have enterpreneurial interests in seeing them developed and implemented. We need democratic mechanisms and not in the form of committees with no decision-making power who are told to go off and think about the ethical issues.

Take an example: At present, the Office of Human Genome Research of the National Institutes of Health, which is being directed by James Watson, has decided to devote between 1 and 3 percent of its large budget to research into the ethical issues raised by the genome project (the project to analyze and map all the human genes, which I discussed in chapter 6). But unless the ethical and social impacts of the kinds of information and technologies the Human Genome Initiative will generate are considered at the point at which scientists decide what scientific and technical research gets done, research into the ethical questions is window dressing.

12

Medical, Legal, and Social Implications of Prenatal Technologies

In the early 1960s it became possible to identify fetuses who are at risk for serious disabilities because they are Rh positive while their mothers are Rh negative. Such women produce antibodies against the Rh antigen in the fetus's blood, and, during successive pregnancies, the antibody concentration can increase sufficiently to endanger the health, or even the life, of the fetus they are carrying. This was the first instance of prenatal diagnosis. The rationale for performing the diagnosis before birth was to enable physicians to be ready to give the baby massive blood transfusions immediately after birth so as to minimize the damage. More recently, physicians have become able to transfuse such a fetus in utero. A few other prenatal diagnostic procedures were developed in the late 1960s and early 1970s.

Since 1972, when *Roe* v. *Wade* made abortion legal at least until the twenty-fourth week of pregnancy, an early diagnosis of fetal health problems has given women the option to abort a fetus that is expected to be born with a disease or disability with which they feel unable to cope. And the availability of prenatal tests has made it possible for women who have reason to believe their prospective children are at risk for a particular, serious health problem to go ahead and become pregnant knowing that they can find out whether the fetus they are carrying is affected with the disease in question. They then have the chance to decide whether to carry the pregnancy to term. Needless to say, the tests cannot guarantee that the fetus does not have some other, unanticipated, health problem. But that is true for any of us and the chance of its happening is small.

By now, quite a number of prenatal tests can be performed (Blatt, 1988), and many more will become available as the project to identify and sequence the human genome gets under way (see chapter 6). Some of them involve simply testing samples of the pregnant woman's blood; others are more invasive. Amniocentesis, one of the more usual procedures, requires that a sample of the amniotic fluid that surrounds the fetus in the womb be withdrawn by means of a hypodermic syringe and needle inserted near the pregnant woman's navel. The amniotic fluid can then be tested directly, or the fetal cells that are suspended in it can be cultured under sterile conditions and tested biochemically or examined under the microscope when a sufficient number of cells have accumulated. Amniocentesis cannot be performed until there is sufficient amniotic fluid, which usually requires waiting until about the sixteenth to eighteenth week of pregnancy.

A still experimental procedure, called chorionic villus sampling (CVS), enables physicians to collect fetal cells considerably earlier in a pregnancy than amniocentesis does. The cells are obtained by inserting a probe through the cervix into the uterus and snipping a small sample from the chorion, which is one of membranes that surrounds the fetus. CVS must be performed during the first trimester, between the eighth and tenth weeks of pregnancy. Another procedure, called fetoscopy, is used rarely. It involves withdrawing samples of blood or other body fluids or tissues from the fetus itself.

Health Risks of Medicalizing Pregnancy and Birth

Feminist scholars have been documenting the transformation of birth and the ways this change has affected our concepts of motherhood (Ehrenreich and English, 1978; Leavitt, 1987; Rothman, 1982, 1989; Wertz and Wertz, 1977). Birth used to be a social event, experienced at home, in which the birthing woman could get advice and support from female relatives and friends who had borne children. If a midwife was present, she usually was from the birthing woman's class and ethnic and racial group and might know her socially and have attended births by her friends and neighbors. Now childbirth takes place in a hospital, where it is made to conform to medical and hospital routines. Judith Walzer Leavitt (1987) has shown that as long as women gave birth at home, they retained considerable control, even when they were attended by male physicians. And given the risks of childbirth during the nineteenth century, many women who could afford it felt more secure in the care of physicians than they did without them. The real change, Leavitt argues, came with the move from the home, which was the birthing woman's turf, to the hospital, where physicians were in charge.

In 1900, about half of U.S. births were attended by midwives; in 1935, only about one-sixth, mostly in the rural South; and by 1972, 99 percent of births were attended by physicians. Fewer than 5 per cent of women had their babies in hospitals in 1900; about half the births took place there in 1940; and essentially all of them did by 1960.

During the same period, many other changes occurred in American society and in medical knowledge and practices. New scientific ideas were formulated about the causes and proper treatment of disease. Industrialization changed patterns of transportation and urbanization, which affected the availability of food and produced changes in diet. People's ways of working and living were transformed. Maternal and infant mortality rates decreased, and life expectancies increased.

The lower rates of maternal and infant deaths and diseases cannot be attributed to any one of these changes and certainly not to the changes in birth practices. Quite the contrary. The shifts from midwives to physicians and from home to hospital births were detrimental to the health of many women and babies, particularly the poorer ones who often ended up with less individualized and expert care than before or with no care at all.

Since the beginning of medical interventions in birth, these have held some risks for women and their babies. From the eighteenth century until the discovery of antibiotics in the late 1930s, childbed fever (also called puerperal fever) took an enormous toll on women's lives and health. It was brought on by physicians with insufficient understanding of their role in spreading bacterial infections.

Similarly, artificial induction and excessive use of forceps damaged infants and their mothers. In 1920, in the first issue of the *American Journal of Obstetrics and Gynecology,* Dr. Charles B. Reed described several methods to induce birth artificially, which, he claimed, were safer than normal birth. In the same issue, Dr. Joseph DeLee advocated the prophylactic (that is, preventive) use of forceps and episiotomy (enlarging the vaginal opening by making a deep cut in the vaginal muscle). He wrote that with proper management and repair, episiotomies produced healthier babies and less debilitated women with "virginal" vaginas, which leads me to conclude that he was primarily concerned with the advantages a woman's husband would derive from the procedure. Dr. DeLee also advocated using morphine and scopolamine. He wrote that women were so frequently injured during childbirth that he "often wondered whether Nature did not deliberately intend women to be used up in the process of reproduction, in a manner analogous to that of the salmon which dies after spawning," a poetic metaphor that may have been rendered more apt by his forceps and scalpel.

By the 1940s and 1950s, hospitals were routinely using hormones to initiate and speed labor, and barbiturates and scopolamine to erase all memory of the birthing experience. Because these drugs impede the higher brain functions, they were said to induce twilight sleep, a condition in which women could "take orders" from their physicians but not "know" what was happening to them or "remember" it. Because the drugs suspended women's capacities to think rationally, they were tied down to keep them from injuring themselves as they thrashed about during labor. A labor room of the period was a superb confirmation of the cultural stereotype of women as irrational creatures who needed knowledgeable and rational men to protect them against their own unreason. There is no better place to go for a description of what this kind of birthing felt like than to Adrienne Rich's (1976) account of giving birth in Boston during the 1950s.

Twilight sleep throughout labor and birth was in wide use until the 1960s. It was discontinued partly because of the opposition of women's health advocates to the unnecessary medicalization of birth, partly because of mounting evidence that it was bad for babies. Fetuses inevitably got their share of the medication given to birthing women, so that the babies were born half-asleep, limp, and often in need of resuscitation. In the 1950s some hospitals began to depend on spinal anaesthesia, which allowed birthing women to remain conscious without feeling pain because it deprived them of all sensations below the waist. As a result, they could not push the baby out, so birthing required the use of forceps, another potential source of damage to newborns and their mothers.

The early 1960s witnessed the thalidomide disaster (Sjöström and Nilsson, 1972). Physicians prescribed this drug to allay nausea and other discomforts some women experience early in pregnancy. Before thalidomide was recalled, its use had resulted in the birth of thousands of limbless infants in Great Britain, Germany and other countries of Western Europe, and Canada. In the United States, we were spared only by the thoroughness and foresight of Dr. Frances E. Kelsey of the federal Food and Drug Administration, who refused to clear this new drug for sale because of insufficient proof of its safety.

During the 1950s and 1960s physicians prescribed the hormones progestin and diethyl stilbestrol (D.E.S.), sometimes routinely, in the mistaken belief that they prevented miscarriages early in pregnancy. Both can harm a fetus. However, because progestin induces excessive growth of the infant's clitoris so that it resembles a penis, the damage was obvious at birth and the use in pregnancy was stopped before long. Unfortunately, physicians did not realize until the late 1970s that D.E.S. could induce a rare form of vaginal cancer and perhaps also testicular cancer and reproductive deficiencies among the daughters and sons pregnant women were carrying at the time they received the drug.

It would be a mistake to believe that all, or perhaps even most, women were unwilling victims of these medical interventions. Many women welcomed the relief offered by drugs, much as they welcomed contraception and baby bottles as respites from the stresses of motherhood. They accepted the reasons physicians gave why drugs and other interventions were necessary as well as physicians' assurances that the interventions were safe and likely to improve birth outcomes. Yet with the best will in the world, physicians cannot foresee the risks of their interventions in pregnancy and birth—risks for women and for our children. At present, physicians seem to feel that it is all right to use the new diagnostic tools and therapies as long as prospective parents have the opportunity to give their informed consent. But what does "informed" mean when applied to new procedures whose benefits and risks cannot be assessed accurately?

Most therapies become established on the basis of custom and professional consensus and are not preceded by rigorous, scientific evaluations of their outcomes. Usually, by the time therapies are tested in scientifically controlled clinical trials, they have been in use for years. Even after the trials are completed, the results are often contested. For this reason many clinicians trust their intuition and experience more than they trust scientific experiments.

Most of the information prospective parents get about the relative merits of different ways to proceed during pregnancy and birth depends on their physician's ideas about the appropriate course. Particularly with new interventions, prospective parents have access to few, if any, other sources of information. Therefore "informed consent" is better than nothing because it at least obliges physicians to try to explain what they plan to do and their reasons, but it serves mainly to provide legal protection for practitioners and hospitals. There is no way people who do not have access to a range of sources of information can make independent judgments and give truly informed consent (Henifin, Hubbard, and Norsigian, 1988).

That is a problem "lay" people always confront when they must make choices about technical matters, be it nuclear energy or prenatal interventions. If there is disagreement among the experts—and there often is—it usually boils down to deciding, on the basis of various criteria, which experts to trust. In the case of nuclear energy, we can gain at least some reassurance from the fact that the experts must live with the outcome. When it comes to interventions in pregnancy, only we and our families have to live with the results, not our physicians.

All tests must be as specific and accurate as possible. That means that there must be a high degree of probability that the condition one intends to test for is the only one being tested, and that the test will not indicate that the condition is present when it is not (false positives), or that it is not present when it is (false negatives). No test satisfies these criteria perfectly, but the better the test, the closer it must come to doing so.

What other risks need women consider? Tests that can be done on samples of blood drawn from the pregnant woman are not likely to impose physical risks because drawing blood is fairly routine. At present, blood samples can be used to measure the level of alpha-feto-protein (AFP), a protein secreted by the fetus at certain stages of development that enters the bloodstream of the pregnant woman. Maternal serum alpha-feto-protein (MSAFP) levels are used to indicate whether the baby is likely to have spina bifida (a malformation of the neural tube) or anencephaly (no brain), both quite rare conditions. MSAFP levels are also now being used as preliminary indications that the baby may have Down syndrome.

Because amniocentesis involves inserting a needle through a pregnant woman's abdominal wall into her uterus, it is more invasive and riskier than blood tests. This is so even when the amniocentesis is done while monitoring the position of the fetus and placenta by means of ultrasound so as not to damage them. If sufficient amniotic fluid has been collected, its AFP content and other chemical properties can be checked. The fetal cells that float in the amniotic fluid can be biochemically tested for specific diseases such as cystic fibrosis, sickle-cell anemia, or Tay-Sachs disease if there is some reason to think the baby might inherit one of them. In order to detect chromosomal abnormalities, such as are present in people who have Down syndrome, it is necessary to culture the fetal cells, which can take two weeks or more. Ultrasound by itself can reveal anatomical malformations of the fetal skeleton, nervous system, kidneys, and other organs.

Ultrasound is said to be safe because no ill effects are seen in newborns and children, but it is not clear how to evaluate this claim. There is no question that at higher levels than those ordinarily used for diagnosis ultrasound damages chromosomes and other intracellular structures and breaks up cells. And, as with other radiation, it is questionable whether there is a threshold level below which ultrasound is absolutely safe and to what extent the effects of successive exposures may be cumulative. Symposia that have reviewed the available evidence have usually ended by warning against the indiscriminate use of ultrasound, and some physicians continue to urge caution (Banta, 1981; Bolsen, 1982).

By now, ultrasound is used so routinely to monitor pregnancy and birth and its effects could be so varied that it will take extremely careful studies, involving large numbers of children over long periods of time, to determine whether there are risks and what they are. While such studies are in progress, prospective parents must rely on physicians' assurances not to worry. Unfortunately we were also told not to worry about x-rays, which eventually were shown to provoke an increase in the incidence of childhood cancers and leukemias. It is always a question of balancing possible risks and benefits. In some situations, the immediate benefits of ultrasound imaging clearly outweigh its possible long-term risks. At other times, the balance of benefits and risks is not so clear, and it is often hard to know where to draw the line. Unfortunately, at present many obstetricians believe that using ultrasound involves no risks.

If it is done by an experienced practitioner, amniocentesis carries a small risk of mechanical injury to the fetus or placenta and a somewhat greater risk of infection. In about one case in three or four hundred, for unknown reasons, amniocentesis results in a spontaneous abortion. CVS may involve less risk of infection than does amniocentesis but a somewhat greater risk of spontaneous abortion. There is also a greater chance of wrong diagnoses with CVS because the cells that are removed from the fetal membranes do not always have the same chromosomal constitution as the fetus itself (Rhoads et al., 1989). However, CVS has the advantage of being performed sufficiently early to permit a first-trimester abortion if the results lead the woman to decide to have one. Amniocentesis, however, necessarily involves a second-trimester abortion, which is more dangerous and psychologically stressful. Risk of spontaneous abortion from fetoscopy is much greater than from amniocentesis or CVS and so are the risks of infection and mechanical damage.

If the only alternative to prenatal testing is not to have children because one or both partners consider the risk of having a child with a particular health problem to be too high, they may be prepared to accept considerable risks from prenatal diagnosis, hoping that they will learn that the fetus does not have the disability in question. Even so, they may have difficulty accepting the possibility of injuring or losing a fetus that would have been healthy.

Because of the fear and costs of malpractice suits, physicians increasingly feel that they need to perform tests so as to be legally covered in case a baby is born with a health problem they could have detected. And because of the responsibilities and costs of raising a child with a serious disability, prospective parents also feel pressure to use the tests. Therefore, as the number of conditions that can be diagnosed before birth increases, more women will have to decide whether to undergo testing. If they do, they will experience the uncertainty of waiting for results, which is often the most difficult part of the procedure. Reports can be slow to come, and the decision whether to abort becomes increasingly difficult as the pregnancy advances (Rothman, 1986). There is no reason to doubt that this kind of stress on a pregnant woman gets communicated to her fetus. Yet threats of legal action against pregnant women as well as physicians increase the pressure to test (Henifin, Hubbard, and Norsigian, 1988).

If it is too soon reliably to evaluate the medical risks of prenatal diagnosis, this is even more true for fetal therapy, which is newer. At this point, fetuses are being treated mostly for hydrocephalus ("water on the brain") and for malformations or malfunctioning of the urinary tracts. These are reasonable conditions to try to remedy during pregnancy because the fetus has a better chance to develop normally if the problem is repaired. However, the risk exists that the problem will be repaired but that the baby will be born with life-threatening disabilities. To date, several fetuses who were treated in utero have been born

with serious disabilities; a few have been normal; still others
have died before or shortly after birth. And, of course, all the
interventions are hazardous for the pregnant woman.

Some Legal Risks

In addition to the medical risks, pregnancy interventions involve dangers to the legal rights of parents and especially of
pregnant women. New choices all too readily become obligations to make the "right" choice by "choosing" the socially approved alternative. And people end up with little, if any, choice
at all.

It stands to reason that as long as pregnant women retain the
right to abortion, they can also refuse prenatal diagnostic and
therapeutic procedures. They can abort a fetus whose health
they have reason to think may be compromised, or they can
allow the pregnancy to continue without intervening. Yet, as we
will see in the next chapter, some legal scholars are challenging
women's right to refuse prenatal diagnosis or treatment, while
upholding our right to abortion (Robertson, 1983). Even without going into that question, legal questions arise during the
third trimester of pregnancy, when the fetus becomes viable
outside the womb. To what extent can a pregnant woman refuse
diagnostic or therapeutic procedures after the fetus reaches
viability when an obstetrician says they have a good chance of
benefiting the health and well-being of her future child? When
therapeutic interventions are available, most pregnant women
yield to the judgment of their physicians. But what if they don't?
If a physician decides that a given test or therapy could benefit
the future child, can the procedure be performed against the
pregnant woman's wishes?

Few experts are qualified to evaluate new procedures. Therefore second opinions are hard to come by. But unless a woman
can find an equally qualified physician to back her up, the physician who purports to be speaking in the interest of the fetus
may well have the last word. If the woman remains adamant in
her refusal, the physician may insist on a legal ruling. Then it is
up to the court (which usually means a male judge) to deter-

mine the appropriateness of proceeding even when the proposed therapy requires giving the woman drugs or submitting her to surgical procedures of which she disapproves. In the 1980s the ground shifted in this area and gave way to an alarming willingness to override women's choices and control over our bodily integrity (Gallagher, 1988).

As recently as 1979, the medical journal *Obstetrics and Gynecology* published two companion articles that supported women's decisions in these situations. The authors—four physicians and an attorney—argued that no matter how much an attending physician disapproves of a woman's decision, she has the right to refuse a Caesarean section, even when the physician is sure that the operation offers the only chance that the fetus will survive. The physicians write (Lieberman et al., 1979): "As far as legal theory is concerned, a person is defined as any being whom the law regards as capable of rights and duties, beginning at the birth of the child until the death of the adult. Most definitions require complete delivery of the newborn from his [*sic*] mother before the status of a legal personality is acquired." They suggest that, when confronted with "maternal unresponsiveness," a physician be allowed to warn the birthing woman that she is committing a felony. But according to them, the physician does not have the duty to perform a Caesarean against the woman's will.

In the accompanying legal analysis, Thomas Shriner (1979) rejects even the physicians' proposed warning. Pointing out that "warnings of felony conjure up the policeman, not the obstetrician," he argues:

> Such warning would almost certainly be false. This results, in part, from the very concept of a person . . . [as] a being capable of rights and duties. A woman is plainly a person; a fetus' status is not so clear. A woman's undoubted personhood carries with it the ancient and well-nigh sacred right that no one may touch her body without her consent. Hence, she may refuse surgery for any reason, or for no reason at all. A fetus, however, for a variety of reasons, may never even be born. . . . [Even if one assumes that a full-term fetus is a person,] the woman's right not to undergo standard, low-risk surgery prevail[s] over the fetus' right to live. . . . [The reason is] biological. The woman can exercise her right by doing nothing; the fetus can exercise its right only if a third person, the obstetrician, performs the affirmative act of cutting open the woman's

body, thereby subjecting himself or herself to criminal liability. The
fetus' chances of prevailing under these circumstances are not very good.

Shriner concludes: "The current law appears to embody the best
general rule. There is no acceptable alternative to requiring the
woman's consent to surgery, and the obstetrician's role must remain one of informing, counselling, and persuading in the
difficult, but fortunately infrequent, situation which Lieberman
et al. present."

However, a year after these articles were published, the
Georgia Supreme Court ordered a Caesarean section for a
woman who refused consent. She was lucky and delivered va-
ginally before the decree could be implemented.

I want to describe an even more worrisome case, reported in
Obstetrics and Gynecology only two years after the articles I just
quoted (Bowes and Selegstad, 1981). A woman arrived at the
hospital of the University of Colorado in Denver near term.
Her amniotic membranes had burst, and she shortly went into
labor. After a while, physicians noticed signs of fetal distress
and advised her to have a Caesarean. She refused "because of
her fear of surgery." A psychiatric consultant who was called in
certified that she was "capable of understanding the circumstances and making a rational decision." At this point the obstetrician in charge summoned the hospital attorney. After
preliminary interviews with the woman and the obstetrician, a
judicial hearing was convened in the hospital, presided over by
a judge of the Denver juvenile court. Court-appointed attorneys
represented "the patient and the unborn infant." (Throughout
the article the fetus is referred to as the "unborn infant" or "unborn child" and the birthing woman as the "patient.") The article reports that "the court found that the unborn baby of the
patient was a dependent and neglected child within the meaning of the Colorado Children's Code. It was then ordered that a
Caesarean section be performed to safeguard the life of the unborn child." Note the authors' use of the passive voice to disavow responsibility. Not "the judge ordered" or even "the court
ordered," but "it was ordered." By whom? God?

Clearly, the medicoethical and legal conclusions reached by
the courts in Georgia and Colorado differed from those ex-

pressed in the articles in *Obstetrics and Gynecology* only a year or two before. The Colorado decision to perform a Caesarean against the "patient's" wishes was based on the state's "*compelling interest to protect the unborn in light of Roe* v. *Wade* and *Doe* v. *Bolton*" (original emphasis). This represents a serious distortion of the *Roe* decision.

Since then several other Caesareans have been performed on the basis of court orders issued against the expressed will of the pregnant woman (Gallagher, 1987; Kolder, Gallagher, and Parsons, 1987). In the most egregious of them a pregnant woman who was dying of cancer was forced by George Washington University Hospital to undergo a Caesarean section against her will and those of her husband, mother, and physicians. The hospital obtained a court order to "deliver" her fetus in the twenty-sixth week of pregnancy. The baby died within two hours of the operation, and the mother died two days later. A physician has testified that the surgery speeded her death. At present, the granting of the court order is under appeal before the highest court in Washington, D.C., and the appeal has been joined by a large number of religious, medical, political, civil liberties, and public-interest organizations. At the same time, the woman's family has filed a suit against the hospital.

There is a real danger that these legal decisions and the arguments put forward to justify them can be broadened to deny women the right to refuse diagnostic or therapeutic procedures intended to improve the health of the fetus at early stages during the pregnancy and especially once the fetus reaches viability (Henifin, 1986; Henifin, Hubbard, and Norsigian, 1988). Such a result would give obstetricians and judges entirely too much control over prospective parents and especially over pregnant women; but more of that in the next chapter.

A Few Social and Economic Considerations

The feminist sociologist Leta S. Hollingsworth wrote in 1916: "An exceedingly important and interesting phase of social control is the control by those in social power over those individuals who alone can bring forth the human young, and thus perpetuate society." She contradicted physicians of her time, who were using the "maternal instinct" to provide naturalistic explanations for women's place in society, and argued that childbearing and child rearing are used as reasons to keep women in line. At present, the steadily expanding repertory of prenatal tests may enable obstetricians, with the help of the courts, to act out once again society's contradictory beliefs that women's reproductive functions are our "natural" and principal calling, yet that they are suffused with pathology and require medical intervention.

Some people counter such fears with the argument that excessive medicalization of pregnancy and birth will be limited by fiscal constraints. Apart from the fact that fiscal limitations are an unfair way to regulate use, the argument does not hold together. It is expensive for a physican or a hospital to acquire the necessary expertise and equipment to offer the most up-to-date scientific and technical innovations. By paying for the services, whether directly or through third parties, patients in fact help the professionals and institutions to amortize their start-up costs.

All the while, these interventions reshape the childbearing experience in ways we often do not even notice. That is what I want us to look at next.

Cost encourages rather than limits use by those who can pay . . . .

13

Prenatal Technologies and the Experience of Childbearing

I now want to explore some implications of the prenatal technologies for the experience of childbearing. Obviously, I do not mean to imply that there is just one kind of childbearing experience. I assume there are as many experiences as there are women bearing children, and that they range from bliss to agony because they depend on the social and personal circumstances of our lives. However, the new technologies raise issues that can affect all these experiences, although they may affect different ones differently.

Childbearing: A Social Construct

Before going further, I want to clear up one point: I am not trying to distinguish between "natural" and "technological" childbearing practices (and by childbearing I mean the entire range of women's activities from conception through birth and lactation). No human pregnancy is simply "natural." Societies define, order, circumscribe, and interpret all our activities and experiences. Just as our sexual practices are socially constructed and not a natural unfolding of inborn instincts, so our ways of structuring and experiencing pregnancy and birth are shaped by society.

Whether a woman goes off to give birth by herself (as !Kung women do in the Kalahari desert), calls in neighboring women and perhaps a midwife (as my grandmother did in her small town in eastern Europe), goes to a lying-in hospital (as I did in

Boston around 1960), or has a lay midwife be with her and her partner at home (as several of my friends have done recently), all these are socially devised ways, approved by one's community, even if sometimes not by the medical profession or the state.

The question we feminists must ask is not which is more "natural" but to what extent different ways of giving birth empower women or, alternatively, decrease our power to structure childbearing around our own needs and those of the people with whom we live.

By now, periodic visualization of the fetus by means of ultrasound is considered routine prenatal care by many physicians in the United States and much of Europe. Indeed, in some places ultrasound visualization is mandatory at least once or twice during pregnancy. Real-time ultrasound recording allows women and their attendants to view the fetus, so to speak, in action. It also tends to reveal its sex. Most prospective parents agree that ultrasound visualization makes the fetus more real, more their baby. However, as we saw in the last chapter, for some women (although for which ones in the United States may depend more on their social and economic circumstances than on their health needs), ultrasound visualization is followed by amniocentesis and the possibility of a second-trimester abortion, so no baby.

These interventions, and indeed the mere fact that they may occur, affect the way we look on our pregnancies (Petchesky, 1987; Rothman, 1986). At the very least, we must decide whether to accept the interventions and how far to take them— something we can still usually do. So you see that pregnancy has become very different from what it was as recently as the 1970s, when once women decided to become pregnant (or to accept an accidental pregnancy), they did not face further decisions about whether to carry the pregnancy to term.

Let me be clear: I completely support every woman's right to decide whether and when to bear a child. She must, therefore, have the right to abort a fetus, whatever her reasons. What is more, on the basis of my own and other people's experiences as pregnancy counselors, I know that the decision to abort need not be traumatic. Whether it is depends on the social context in which it is made and implemented. But it is one thing to terminate a pregnancy when we don't want to be pregnant and quite another to want a baby but to decide to abort the particular fetus we are carrying in the hope of coming up with a "better" one next time (Asch, 1988; Rothman, 1986).

Research studies and personal accounts are beginning to document women's mixed responses to prenatal diagnosis and the necessity to decide whether to terminate a wanted pregnancy because the baby may be abnormal (Blumberg, Golbus, and Hansen, 1975; Kenen, 1981; Neilsen, 1981; Rapp, 1984; Rothman, 1986). The abortion itself may not be so bad once the decision is made, but some of my friends have told me how much they have hated the two or three weeks of waiting while the fetal cells were being cultured and tested, knowing that they might end up deciding not to continue the pregnancy. This state of uncertainty can last until the twentieth week—halfway through the pregnancy and several weeks after most women begin to feel the fetus move. By that time, many women look on it very much as their baby.

Ultrasound and amniocentesis confront women with a contradiction: Ultrasound makes the fetus more real and more our baby, while the possibility of following it with an abortion makes us want to keep our emotional distance in case we will not end up with a baby after all.

The specific problems are somewhat different with CVS, which is still considered experimental but is likely to be generally available soon. As we saw in the last chapter, CVS results somewhat more frequently in wrong diagnoses (both positive and negative) than does amniocentesis and in a slightly higher rate of spontaneous abortions. However, it lets women who have reason to fear for the health of their fetus have tests early enough to allow for a first-trimester abortion if they decide to terminate the pregnancy.

When even easier methods than we now have become available to examine a fetus early in pregnancy, the pressure to screen fetuses will increase. Once scientists develop a way to identify cells of fetal origin in the bloodstream of pregnant women, which is likely to happen before long, fetal screening could become a routine part of prenatal care.

The Question of Choice

The means to "choose" the kind of baby a parent will accept bring their own problems no matter at what point in the pregnancy they can be used. To sort these problems out, we need to put this new kind of decision into historical perspective.

In most cultures women have exercised a measure of choice about procreation by practicing some form of birth control—from contraceptive and abortifacient herbs and barrier methods to infanticide. But, until recent times, unwanted pregnancies, hazards of childbearing, and high rates of infant and early childhood mortality have made women's reproductive lives largely a matter of chance. Only during the last century have contraception and abortion gradually become sufficiently available, accepted, and reliable that socially and economically privileged women expect to be able to choose whether and when to become pregnant. At the same time, in the technologically developed, affluent countries, maternal and child health have improved sufficiently for us to be able to assume that the children we bear will become adults.

Full of this new confidence, the more privileged of us seem to have glided over into the illusion that we can control not only whether and when to have children, but the kind and quality of children we will "choose" to have. Barbara Katz Rothman (1984, 1989) points out that in this consumer society people tend to look on children as products that they can or cannot "afford." And, by that way of reckoning, it is realistic to look on the prospect of raising a child with a serious disability as beyond our means. In our economy, raising children is expensive. And in the United States, which is the only industrialized country besides South Africa without universal health insurance, the expense of bringing up a child with a disability can be overwhelming. Indeed, Americans often meet the challenge of disability with litigation as a way to increase their financial resources.

Recent medical and social practices have made it possible to commodify procreation all along the line, with eggs, sperm, embryos, "surrogate mothers," and babies available for a price. Some babies who are up for adoption are for sale outright, although this is illegal. Legal adoption agencies prohibit cash payments, but because they are at pains to establish the financial "soundness" of prospective parents, economic status is clearly important for being considered a fit adoptive parent. Once procreation is a form of commodity production, it is an easy step to require quality control. And at this point genetic screening, ultrasound visualization, and the other prenatal tests come in.

Yet all these tests carry a price tag, and many of them are scarce resources. They are not, and indeed cannot be, available to everyone. While affluent women come to view the new tests and other techniques as liberating advances that improve our lives, economically less privileged and socially more defenseless women continue to be deprived of the ability to procreate (Nsiah-Jefferson, 1988). Forced sterilizations still happen, often by means of hysterectomy (McDonald, 1981; Rodriguez-Trias, 1982). Less extreme measures also deny choice. For example, legal and financial restrictions that limit access to abortion force pregnant women who want an abortion but cannot afford it to "choose" a hysterectomy, which is covered by social insurance in all states except Arizona, which has never had Medicaid.

The consumerist way of looking at procreation creates the illusion that at least those of us who can afford prenatal tests have the choice to have healthy babies. But that choice is a mirage because it exists only in a few circumscribed situations. Each test can provide information only about a specific disability. Therefore tests help parents who have reason to worry that their child will be born with a particular disability. They cannot guarantee a healthy baby. But before I say more about the limits of our ability to predict the health of our future children, I want to raise another issue.

Trade-Offs of Scientific Progress

Feminists have often portrayed medical interventions in pregnancy as part of an attempt by men to control women's capacity to bear children. And although I agree with much in this analysis, I am bothered by the way it downgrades, and sometimes romanticizes, the pain and travail childbearing has meant for many women. Bearing and rearing children is difficult under the conditions in which most women live. Judith Walzer Leavitt (1987) amply illustrates this in her book *Brought to Bed,* in which she describes the history of childbearing in America. The literary historian Ruth Perry (1979) has calculated that in eighteenth-century England a married woman who "delivered six children (a not unusual number) . . . had at least a ten per cent chance of dying, and probably a much higher one." It is small wonder that women have welcomed anesthetics and the other interventions physicians introduced, hoping that these would lessen the danger and pain of pregnancy and birth.

True, the medical "improvements" took their toll. As we have seen, they brought their own dangers and gave physicians altogether too much authority and control over the ways women experience pregnancy and birth as well as over child rearing. Yet, given the limited choices most women have had and the very real risks of childbearing, it is not surprising that upper-class women in the nineteenth century opted for the benefits their physicians promised them.

Now once again we are at a point where many women believe that the new interventions in pregnancy are increasing pro-creative choice and improving our lives as well as those of our families. As the new technologies become part of routine pregnancy management, the experiences of past generations of women are erased, and contemporary women find it impossible to imagine how they could live without technologies that women in the past did not miss and often were better off doing without.

Take the pill as an example. Many women who became heterosexually active in the early 1960s seem to think that birth control was rare, if not unknown, for women earlier in the century. Yet in the pre-pill era, many women planned their pregnancies as successfully as women who use the newer products do now. True, condoms and diaphragms involve problems and inconveniences, but from a health viewpoint they are safer than the newer methods. And these days, when protection against AIDS should be on all our minds, a shift back to condoms, used together with spermicides, is widely recommended.

Another example is pregnancies of older women. In this context "older" denotes younger and younger women as time goes on. It used to be over forty; in this country, it is now over thirty-five; I suppose thirty-two and thirty are next. This change seems strange to me, who had my children when I was between thirty-five and forty. I did not think there was anything to worry about because my partner and I were in good health, and when I mentioned my age to the obstetrician, he agreed with this assessment.

Now women that age tell me that if it were not for prenatal diagnosis, they would not dare to have a child. Indeed, because it has been hammered into the present generation of "older" women that the risk of bearing a child with a chromosomal abnormality, such as Down syndrome, increases dramatically after the early thirties, few will just hope for the best when a test can reassure them while they are pregnant. And because all abnormalities are rare, it usually does just that. This reassuring feature is what makes for the popularity of the tests: More often than not they show that there is no problem.

But most of us would not need that reassurance if it were not for the prevalent emphasis on risks. The reasons why chromosomal abnormalities occur when they do are far from clear, and there has not been a great deal of epidemiological research on them. What are the environmental, occupational, and socioeconomic influences? How relevant are the health histories of both partners? And so forth. In the midst of such uncertainties, the prospective mother's age is the only factor we are urged to consider, even though either prospective parent can contribute the extra chromosome responsible for Down syndrome.

I have a not so sci-fi fantasy in which a woman ten years hence will tell me that she could not possibly risk having a child by "in-body fertilization." She will say that only the availability of in vitro fertilization makes childbearing possible. In my fantasy, it will by then be standard practice to fertilize eggs in vitro and allow the embryo to go through the first few cell divisions until it contains six or eight cells. At this point two of its cells will be removed, cultured, and put through a battery of tests, while the rest of the embryo will be frozen and placed in cold storage. Only if the tests are satisfactory will the embryo be thawed and implanted in the prospective mother's, or some other carrier's womb. Otherwise, it will be discarded or used for research.

How will I explain to this woman why I am troubled by this,

by then routine, way of producing babies? We will live in different worlds. I in one in which I continue to look upon childbearing as a healthy, normal function that can sometimes go wrong but usually doesn't. Therefore I will want to interfere as little as possible with the delicate, complicated processes of fertilization and embryonic development. She will live in a world in which the ability to plan procreation means using all available medical techniques to try to avoid the possibility of biological malfunctioning. I will tell her that the manipulations entail unknown and unpredictable risks and that they cannot assure her of having a healthy baby. She will tell me that I am opting for ignorance and stemming progress. But what worries me most is that at that point "in-body fertilization" will not only be considered old-fashioned and quaint, but foolhardy, unhealthful, and unsafe. It will seem that way to the scientists and physicians who pioneer the "improvements" and to the women who "choose" to have their babies the new way.

Nowadays, some women over thirty-five refuse prenatal diagnosis and other, usually poor or Third World women, do not even know it exists. In reality, in this country only a minority of women are being screened. But this is because access to costly resources is uneven, not because physicians are cautious about exposing women to these techniques. In fact, to most physicians being cautious means using all available technology, however little they can yet know about its long-term consequences. By using every test they can avoid the potential legal ramifications of failing to alert future parents to the possibility of an inborn "defect."

I do not want to portray physicians as callous technocrats who think only about possible malpractice suits. Nor do I want to portray women as unwilling victims of scientists and physicians, quite a number of whom, incidentally, by now are women. When women contemplate childbearing, they try to strike the best bargain they can in a society that offers little support for this important social activity. And sometimes technological interventions seem to offer a measure of security from unexpected mishaps. The problem is not the technology itself but the fact that it generates the expectation that it is up to prospective parents, and especially mothers, to do away with disabilities by not bearing children who might have one.

Disabilities: A Social Problem

What shall be called a defect or disability and for how many and what kind shall a fetus be aborted or treated in the womb? Down syndrome? Spina bifida? Wrong sex? These questions are complicated by the fact that, for most inborn disabilities, no one can predict how serious the "defect" will be and just how it will express itself—in other words, how much of a health or social problem it will be.

To some people, and in some circumstances, the prospect of having a child with Down syndrome, no matter how mild, seems intolerable. Before there were tests, women like myself just hoped for the best when we decided to have children after thirty-five. But now that tests exist, many women who have access to them have them as a matter of course. At the same time, many people in the United States reject prenatal testing for purposes of sex selection, although it is done widely in India and China. Many feminists argue against sex selection because they expect it will most often be used to choose boys; but some feminists are for it because it makes it possible for women to choose to have only daughters if that is what they want.

I have problems with such so-called choices. Bearing and raising children is intrinsically unpredictable and knowing a person's sex tells us little about them. With all the prenatal tests in the world, we cannot know what our children will be like, whether they will be healthy and able-bodied and remain so, and what sorts of people they will be when they grow up. We have the best chance of successful parenthood if we are prepared to accept our children, whoever they are, and do the best we can to help them accept themselves and, hopefully, us too. People with disabilities have begun to speak about these issues. I agree with them when they say that all children should be welcome and that we are being short-sighted to think that we can circumvent the uncertainties of procreation by aborting "defective" or "wrong" fetuses. Sparing no expense to develop techniques for diagnosing disabilities prenatally, so as to prevent the birth of children who have them, accentuates the stigma to which people with disabilities, as well as their families, are exposed.

Another, rather different, issue we must be aware of is that

the increasing emphasis on prenatal testing reinforces this society's unfortunate tendency to individualize people's problems.
Yet disability cannot be dealt with properly as long as it is considered a personal problem. Parents cannot possibly provide on
their own for a child who may outlive them by decades. The
logical solution: Don't have one! Logical, maybe, but neither
humane nor realistic. Disability-rights advocates point out that
usually the disability is not the main problem. What makes it
burdensome is how people are treated because they have it.
And as I have said before, many (probably most) disabilities
cannot be predicted unless we were to test every embryo or
fetus for all conceivable disabilities—an exceedingly cumber-
some and expensive process with little benefit to show for it.
What is more, the incidence of disabilities resulting from accidents or exposure to chemicals or radiation is considerable and
likely to increase, rather than decrease, in the future. It would
be better to regard mental and physical disabilities as social, not
personal, issues. Many disabilities—whether inborn or acquired
later in life—are the result of social circumstances: accidents,
inadequate living conditions, chronic poisoning by heavy metals
or drugs, and so forth. They cannot be dealt with by victim-
blaming individualizations; to prevent them requires social
measures.

As the world around us becomes increasingly hazardous and
threatens us and our children with social disintegration, pollution, accidents, and nuclear catastrophe, it seems as though we
seek shelter among the hazards we are told lurk within us, in
the illusion that we may have at least some control over them.
And so we applaud the scientists and physicians who tell us that
our problems lie in our genes or our womb and who propose
technological solutions for them.

I remember a news story that ran in the *Boston Globe* under
the headline "Some Schizophrenia Linked to Prenatal Changes
in Brain Cells" (Nelson, 1983). It started with the portentous
words: "The devastating mental disorder of paranoid schizophrenia seems to have roots in the womb." The rest of the story
showed nothing of the sort. Rather, on the basis of the flimsiest
evidence gathered by examining the brains of "10 deceased
schizophrenics ages 25 to 67" and "eight nonpsychotic subjects

used as controls," two researchers decided that schizophrenia has its beginnings in "the first few months of pregnancy" and suggested that visible "abnormalities [in the brain] some day may allow doctors to identify children who have a high risk of becoming paranoid schizophrenics." This is but one of many false messages women get that our children's troubles originate in our womb if not in our genes.

As I have said before, prenatal diagnosis can help the relatively small number of women who have reason to think their future children are at risk for a specific disease, but most disabilities are unexpected. Yet now that some disabilities can be detected and the fetus aborted or, in rare instances, treated in the womb, people are beginning to feel that if parents bear a child with a disability, it is because they or their physicians were not sufficiently foresightful.

Parents are suing physicians, arguing they should have been forcefully warned about possible risks of disability and told about all available means of prenatal diagnosis (Henifin, 1986). And a child who is born with a health problem that might have been detected and improved prenatally may be able to sue the mother if she refuses to be tested while pregnant (Henifin, Hubbard, and Norsigian, 1988). Not only that. Some attorneys have even suggested that the state should be able to mandate prenatal screening "with criminal penalties for the woman who fails to obtain it" (Robertson, 1983).

Fetal Rights

In 1971, Bentley Glass, the retiring president of the American Association for the Advancement of Science, wrote: "In a world where each pair must be limited, on the average, to two offspring and no more, the right that must become paramount is . . . the right of every child to be born with a sound physical and mental constitution, based on a sound genotype. No parents will in that future time have a right to burden society with a malformed or a mentally incompetent child."

More recently, the theologian Joseph Fletcher (1980) has written that "we ought to recognize that children are often

abused preconceptively and prenatally—not only by their mothers drinking alcohol, smoking, and using drugs nonmedicinally
but also by their *knowingly* passing on or risking passing on genetic diseases" (original emphasis). This language of "rights" of
the unborn immediately translates into obligations of the born,
and especially of women.

These obligations become explicit in the writings of Margery
Shaw, an attorney and physician. Reviewing what she calls "prenatal torts," Shaw (1980) argues as follows:

> Once a pregnant woman has abandoned her right to abort and has de-
> cided to carry her fetus to term, she incurs a "conditional prospective
> liability" for negligent acts toward her fetus if it should be born alive.
> These acts could be considered negligent fetal abuse resulting in an in-
> jured child. A decision to carry a genetically defective fetus to term
> would be an example. Abuse of alcohol or drugs during pregnancy
> could lead to fetal alcohol syndrome or drug addiction in the infant, re-
> sulting in an assertion that he [*sic*] had been harmed by his mother's acts.
> Withholding of necessary prenatal care, improper nutrition, exposure
> to mutagens and teratogens, or even exposure to the mother's defective
> intrauterine environment caused by her genotype . . . could all result in
> an injured infant who might claim that his right to be born physically
> and mentally sound had been invaded.

What *right* to be born physically and mentally sound? Who has
that kind of right and who guarantees it? Shaw goes on to urge
that "courts and legislatures . . . should . . . take all reasonable
steps to insure that fetuses destined to be born alive are not
handicapped mentally and physically by the negligent acts or
omissions of others."

In this argument, Shaw assumes not only that a fetus has
rights (a hotly debated assumption) but that its rights are different, and indeed opposed to, those of the woman whose body
keeps it alive and who will most likely to be the person who
cares for it once it is born. What is more, she places the burden
of implementing these so-called rights of fetuses squarely on
the shoulders of individual women. Nowhere does Shaw suggest that the "reasonable steps" courts and legislatures should
take include making sure that women have access to good nutrition, housing, education, and employment so that they are
able to secure a fetus its "right" to proper nutrition and avoid

its being exposed to mutagens and teratogens. Her language of "rights" does not advocate the kinds of improvements that would benefit women, children, and everyone. It is a language of social control.

Such control is advocated explicitly by John Robertson (1983), professor of law at the University of Texas (the same faculty on which Shaw teaches). His basic proposition is this:

> The mother has, if she conceives and chooses not to abort, a legal and moral duty to bring the child into the world as healthy as is reasonably possible. She has a duty to avoid actions or omissions that will damage the fetus. . . . In terms of fetal rights, a fetus has no right to be conceived—or, once conceived, to be carried to viability. But once the mother decides not to terminate the pregnancy, the viable fetus acquires rights to have the mother conduct her life in ways that will not injure it.

Because the fetus has such rights, "laws that prohibited pregnant women from obtaining or using alcohol, tobacco, or drugs likely to damage the fetus would be constitutional," and "statutes excluding pregnant women from workplaces inimical to fetal health . . . would be valid." This argument leads Robertson even further:

> The behavioral restrictions on pregnant women and the arguments for mandating fetal therapy and prenatal screening illustrate an important limit on a woman's freedom to control her body during pregnancy. She is free not to conceive, and free to abort after conception and before viability. But once she chooses to carry the child to term, she acquires obligations to assure its wellbeing. These obligations may require her to avoid work, recreation, and medical care choices that are hazardous to the fetus. They also obligate her to preserve her health for the fetus' sake or even allow established therapies to be performed on an affected fetus. Finally, they require that she undergo prenatal screening where there is reason to believe that this screening may identify congenital defects correctable with available therapies.

This analysis gets women into an awkward predicament, although one with a long history. While he was enunciating these legal principles, Professor Robertson was a member of a panel that proposed a model statute to guarantee a person's right to refuse treatment. The statute begins with the following proposition: "A competent person has the right to refuse any medical procedure or treatment" (Legal Advisors Committee, Concern for Dying, 1983). Yet we have just seen that Robertson argues that a woman does not have this right if she becomes pregnant and decides to carry the fetus to term. In that case, she comes entirely under the control of physicians and judges, suggesting that a willingly (or enthusiastically) pregnant woman is not a "competent person."

By defining pregnancy as a conflict of rights between a woman and her fetus, attorneys and judges have injected themselves into the experience of pregnancy, where they see themselves as advocates for the fetus. Judging by other precedents, we can see how this new mechanism of social control could be used against women not only when we are pregnant. It could be expanded to cover every woman of childbearing age by invoking "rights" not just of the fetus she is carrying but of a "potential" fetus—one she may carry at some future date. This expansion shades over into the concept of the "potential" pregnancy I referred to in chapter 2, which has been used to bar women from more prestigious and better-paid jobs than they now have in some male-dominated industries (Stellman and Henifin, 1982).

To present pregnancy as a conflict of rights is even more inappropriate than to regard it as a disease. The disease metaphor is wrong because it turns the special needs of some of us into the norm for us all. The rights metaphor misrepresents most women's experience of a wanted or accepted pregnancy even more than defining pregnancy as a disability does. Yet both Shaw's and Robertson's arguments refer specifically to women who expect to carry their pregnancies to term. A wanted or accepted fetus is part of a pregnant woman's body. For this reason, contrary to Robertson's statement, her decision to carry it to term is not binding. As long as the 1973 Supreme Court decision in *Roe* v. *Wade* stands, she can change her mind and terminate the pregnancy at any point until viability and in some states until birth.

It is in the interest of the well-being of women and children that physicians' judgments not acquire the force of law. Informed consent laws mean that a physician can suggest, advise, and urge treatment, but he or she must not be able to force treatment on unconsenting people. Fetuses cannot consent to tests or treatments. So who speaks for the fetus? Is a judge of the juvenile court, who is called in for the occasion, more appropriate than the woman whose body sustains the fetus and who will be physically, emotionally, and economically affected by whatever is done? If a mother refuses to save her child's life by donating one of her kidneys, no one can force her to do it. What warped logic enables a physician, supported by a judge, to cut her open or penetrate her body with a needle or force her to take medication for the presumed benefit of the fetus she carries inside her?

Fetal Therapy

As we saw in the preceding chapter, the present status of fetal therapy is equivocal. After an initial rush of operations on fetuses that were hailed as breakthroughs in news and feature articles, some of the physicians who pioneered fetal interventions warned to go slow (Harrison et al, 1982). Read cynically, the warning can be taken to mean: "We have begun to gain experience using these procedures at a few of the most prestigious teaching hospitals. Let us do them and don't get into the act!" These same physicians are far from restrained in their other writings about "the fetus." They wax eloquent, and even poetic, over the prospect of treating fetuses. For example, Michael Harrison (1982) writes in a review entitled "Unborn: Historical Perspective of the Fetus as Patient":

> The fetus could not be taken seriously as long as he [*sic*] remained a medical recluse in an opaque womb; and it was not until the last half of this century that the prying eye of the ultrasonogram [that is, ultrasound visualization] rendered the once opaque womb transparent, stripping the veil of mystery from the dark inner sanctum, and letting the light of scientific observation fall on the shy and secretive fetus. . . . Sonography can accurately delineate normal and abnormal fetal anatomy with astounding detail. It can produce not only static images of intact fetuses, but real-time "live" moving pictures. And, unlike all previous techniques, ultrasonic imaging appears to have no harmful effects on mother or fetus. The sonographic voyeur, spying on the unwary fetus, finds him or her a surprisingly active little creatures, and not at all the passive parasite we had imagined.

Who is "we"? Surely not women who have been awakened by the painful kicks of a fetus! Harrison concludes:

> The fetus has come a long way—from biblical "seed" and mystical "homunculus" to an individual with medical problems that can be diagnosed and treated, that is, a patient. Although he [sic] cannot make an appointment and seldom even complains, this patient will at all times need a physician. . . . Treatment of the unborn has had a long and painstaking gestation; the date of confinement is still questionable and viability uncertain. But there is promise that the fetus may become a "born again" patient.

Frederic Frigoletto, chief of what used to be obstetrics but is now called "maternal and fetal medicine" at Boston's Brigham and Women's Hospital and another pioneer in fetal therapy, is quoted as follows in *Patient Care*, a magazine for physicians (Labson, 1983):

> Real-time ultrasound—which is now widely available—allows us to develop a composite picture of [the] fetal state; it's almost like going to a nursery school to watch [the] behavior of 3-year-olds. Eventually we may be able to establish normative behavior for the fetus at various gestational stages. That will help us identify abnormal fetal development, perhaps early enough to be able to correct the environment to treat the fetus in utero.

Considering the personal and social problems that have been created when scientists have tried to establish norms, such as the IQ, for children and grownups, I shudder at the prospect of physicians coming up with norms for "fetal behavior." And remember that the "environment" to which Dr. Frigoletto refers happens to be the body of a woman.

So the fetus is on its way to being a person by virtue of becoming a patient with its own legal rights to medical treatment. As we have seen, once the fetus is considered a person, pregnant women may lose *their* right to refuse treatment by becoming no more than the maternal environment that must be manipulated for the fetus's benefit. The same Dr. Harrison, in a medical article he wrote with colleagues, described moving a pregnant woman as "transporting the fetus in situ" (Harrison, Golbus, and Filly, 1981), *in situ* being scientific parlance for "in place." It is not unusual to find pregnant women referred to as the "maternal environment," but now even that term directs too much attention to them. They are becoming "the fetus in situ," the vessel that holds the fetus, that ideal patient who does not protest or talk back.

And What About Women?

Much as this way of viewing pregnancy insults women, the issue to which I want to come back is that many—perhaps most—pregnant women will feel obligated to accept these intrusions and may even do so gratefully. As long as disability is regarded as a personal failure and parents (especially mothers) feel in some sense responsible, as well as ashamed, if their child is born with a disability, pregnant women will hail medical interventions that promise to lessen the likelihood of its happening to them. The very availability of the new techniques, however untested they may be, increases women's isolation by playing on our sense of personal responsibility to produce healthy children and on our fears and guilt if we should fail to do so.

But who is to say what "healthy" means in the face of an ever-lengthening list of diagnosable "defects" and, lately, even of "predispositions" or "tendencies" to develop them? The Human Genome Initiative will produce a raft of new diagnostic tests long before there will be relevant therapies. Add to that the fact that pharmaceutical companies always find it more profitable to market ways to diagnose and screen healthy people than to develop therapeutic measures because relatively few people need therapy. So how should we relate to the ever-increasing number of genetic and metabolic tests that will be done prenatally?

Compare the rare and almost surreal genetic threats the genome project will uncover with the starkly real threats to health reported daily in the press. For example, in July 1988 the Agency for Toxic Substances and the Disease Registry of the Department of Health and Human Services announced that in the United States an estimated four hundred thousand fetuses a year are exposed to harm because of the lead poisoning of their mothers (*Boston Globe,* 1988). This is only one of many preventable risks pregnant women and fetuses run by reason of economic and social neglect.

Women are likely to accept untested (or insufficiently tested) technological interventions in their pregnancies because it is becoming more and more difficult to be a responsible childbearer and mother. We get lots of "expert" advice, but little comradely support. Unless our social supports improve, women who can afford it will feel driven to follow every new will-o'-the-wisp that promises to lessen our sense that any problems we encounter in our childbearing are our fault. Meanwhile, women who do not have the economic or social support that make it possible for them to experience a healthful pregnancy will be blamed for "their failures."

14

Who Should and Who Should Not Inhabit the World?

Political agitation and education during the past few decades have made most people aware of what constitutes discrimination against blacks and other racial and ethnic minorities and against women. And legal and social measures have been enacted to begin to counter such discrimination. Where people with disabilities are concerned, our level of awareness is low, and the measures that exist are enforced haphazardly. Yet people with disabilities and disability-rights advocates have stressed again and again that it is often far easier to cope with the physical aspects of a disability than with the discrimination and oppression they encounter because of it (Asch, 1988; Asch and Fine, 1988). People shun persons who have disabilities and isolate them so they will not have to see them. They fear them as though the disability were contagious. And it is, in the sense that it forces us to face our own vulnerability.

Most of us would be horrified if a scientist offered to develop a test to diagnose skin color prenatally so as to enable racially mixed people (which means essentially everyone who is considered black and many of those considered white in the Americas) to have light-skinned children. And if the scientist explained that because it is difficult to grow up black in America, he or she wanted to spare people suffering because of the color of their skin, we would counter that it is irresponsible to use scientific means to reinforce racial prejudices. Yet we see nothing wrong, and indeed hail as progress, tests that enable us to try to avoid having children who have disabilities or are said to have a tendency to acquire a specific disease or disability later in life.

The scientists and physicians who develop and implement

these tests believe they are reducing human suffering. This justification seems more appropriate for speed limits, seat-belt laws, and laws to further occupational safety and health than for tests to avoid the existence of certain kinds of people. When it comes to women or to racial or ethnic groups, we insist that it is discriminatory to judge individuals on the basis of their group affiliation. But we lump people with disabilities as though all disabilities were the same and always devastating and as though all people who have one were alike.

Health and physical prowess are poor criteria of human worth. Many of us know people with a disease or disability whom we value highly and so-called healthy people whom we could readily do without. It is fortunate for human variety and variability that most of us are not called on to make such judgments, much less to implement them.

It is not new for people to view disability as a form of pollution, evidence of sin. Disability has been considered divine punishment or, alternatively, the result of witches' spells. In our scientific and medical era we look to heredity for explanations unless there is an obvious external cause, such as an accident or infectious disease. Nowadays, even if an infection can explain the disability, scientists have begun to suggest that our genes might have made us unusually susceptible to it.

In a sense, hereditary disabilities are contagious because they can be passed from one generation to the next. For this reason, well before there was a science of genetics, scientists proposed eugenic measures to stem the perpetuation of "defects."

The Rise of Eugenics in Britain and the United States

Eugenics met its apotheosis under the Nazis, which is why many Germans oppose genetic testing and gene therapy and their use is being hotly debated in the parliament. Germans tend to understand better than people in other countries what can happen when the concern that people with disabilities will become social and economic burdens or that they will lead to a deterioration of the race begins to dictate so-called preventive health policies. They are aware that scientists and physicians were the ones who developed the Nazi policies of "selection and eradication" (*Auslese und Ausmerze*) and who oversaw their execution. What happened under the Nazis has been largely misrepresented and misinterpreted in this country, as well as among Nazi apologists in Germany. To make what happened clearer, I shall briefly review the scientific underpinnings of the Nazi extermination program, which are obscured when these practices are treated as though they were incomprehensible aberrations without historical roots or meaning—a holocaust.

German eugenics, the attempt to improve the German race, or *Volk*, by ridding it of inferior and foreign elements, was based on arguments and policies developed largely in Great Britain and the United States during the latter part of the nineteenth and the beginning of the twentieth centuries. (In what follows I shall not translate the German word *Volk* because it has no English equivalent. The closest is "people," singular, used as a collective noun, as in "the German people *is* patriotic." But "people," singular, does not convey the collectivity of *Volk* because to us "people" means individuals. Therefore, we would ordinarily phrase my example, "the German people *are* patriotic.")

The term *eugenics* is derived from the Greek word for "well born." It was coined in 1883 by Francis Galton, cousin of Charles Darwin, as "a brief word to express the science of improving the stock, which is by no means confined to questions of judicious mating, but which, especially in the case of man [sic], takes cognizance of all the influences that tend in however remote a degree to give the more suitable races or strains of blood a better chance of prevailing speedily over the less suitable than they otherwise would have had" (pp. 24–25). Galton later helped found the English Eugenics Education Society and eventually became its honorary president.

British eugenics counted among its supporters many distinguished biologists and social scientists. Even as late as 1941, while the Nazis were implementing their eugenic extermination program, the distinguished biologist Julian Huxley (1941)— brother of Aldous—opened a semipopular article entitled "The Vital Importance of Eugenics" with the words: "Eugenics is running the usual course of many new ideas. It has ceased to be regarded as a fad, is now receiving serious study, and in the

near future will be regarded as an urgent practical problem."
In the article, he argues that it is crucial for society "to ensure
that mental defectives [*sic*] shall not have children" and defines
as mentally defective "someone with such a feeble mind that he
cannot support himself or look after himself unaided." (Notice
the mix of eugenics and economics.) He says that he refuses to
enter into the argument over whether such "racial degeneration" should be forestalled by "prohibition of marriage" or
"segregation in institutions" combined with "sterilization for
those who are at large." He states as fact that most "mental defects" are hereditary and suggests that it would therefore be
better if one could "discover how to diagnose the carriers of the
defect" who are "apparently normal." "If these could but be detected, and then discouraged *or prevented* from reproducing,
mental defects could very speedily be reduced to negligible
proportions among our population" (my emphasis). It is shock-
ing that at a time when the Nazi program of eugenic sterilization
and euthanasia was in full force across the Channel, Huxley expressed regret that it was "at the moment very difficult to en-
visage methods for putting even a limited constructive program
[of eugenics] into effect" and complained that "that is due as
much to difficulties in our present socioeconomic organization
as to our ignorance of human heredity, and most of all to the
absence of a eugenic sense in the public at large."

The American eugenics movement built on Galton and at-
tained its greatest influence between about 1905 and 1935. An
underlying concern of the eugenicists is expressed in a statement by Lewis Terman (1924), one of the chief engineers of
I.Q. testing: "The fecundity of the family stocks from which
our most gifted children come appears to be definitely on the
wane. . . . It has been figured that if the present differential
birth rate continues 1,000 Harvard graduates will, at the end of
200 years, have but 56 descendants, while in the same period,
1,000 S. Italians will have multiplied to 100,000." To cope with
this dire eventuality, eugenics programs had two prongs: "positive eugenics"—encouraging the "fit" (read "well-to-do") to
have lots of children—and "negative eugenics"—preventing
the "unfit" (defined to include people suffering from so-called
insanity, epilepsy, alcoholism, pauperism, criminality, sexual

perversion, drug abuse, and especially feeble-mindedness) from having any.

Many distinguished American geneticists supported eugenics, but none was more active in promoting it than Charles Davenport, who, after holding faculty appointments at Harvard and the University of Chicago, in 1904 became director of the "station for the experimental study of evolution," which he persuaded the Carnegie Institution of Washington to set up in Cold Spring Harbor on Long Island. His goal was to collect large amounts of data on human inheritance and store them in a central office. In 1910, he managed to persuade the heiress to the Harriman railroad fortune to fund the Eugenics Record Office at Cold Spring Harbor, for which he got additional money from John D. Rockefeller, Jr. He appointed Harry W. Laughlin, a Princeton Ph.D., as superintendent and recruited a staff of young graduates from Radcliffe, Vassar, Cornell, Harvard, and other elite institutions as fieldworkers to accumulate interview data about a large number of so-called mental and social defectives. The office and its staff became major resources for promoting the two legislative programs that formed the backbone of U.S. eugenics: involuntary-sterilization laws and the Immigration Restriction Act of 1924.

The first sterilization law was enacted in Indiana in 1907, and by 1931 some thirty states had compulsory-sterilization laws on their books. Aimed in general at the insane and "feeble-minded" (broadly interpreted to include many recent immigrants and other people who did badly on I.Q. tests because they were functionally illiterate or barely spoke English), these laws often extended to so-called sexual perverts, drug fiends, drunkards, epileptics, and "other diseased and degenerate persons" (Ludmerer, 1972). Although most of these laws were not enforced, by January 1935 some twenty thousand people in the United States had been forcibly sterilized, nearly half of them in California. Indeed, the California law was not repealed until 1980 and eugenic-sterilization laws are still on the books in about twenty states.

The eugenic intent of the Immigration Restriction Act of 1924 was equally explicit. It was designed to decrease the proportion of poor immigrants from southern and eastern Europe so as to give predominance to Americans of British and north European descent. This goal was accomplished by restricting the number of immigrants allowed into the United States from any one country in each calendar year to at most 2 percent of U.S. residents who had been born in that country as listed in the Census of 1890 (so, thirty-four years earlier). The date 1890 was chosen because it established as a baseline the ethnic composition of the U.S. population prior to the major immigrations from eastern and southern Europe, which began in the 1890s. Laughlin of the Eugenics Record Office was one of the most important lobbyists and witnesses at the Congressional hearings that preceded passage of the Immigration Restriction Act and was appointed "expert eugenical agent" of the House Committee on Immigration and Naturalization (Kevles, 1985).

Racial Hygiene in Germany

What was called eugenics in the United States and Britain came to be known as racial hygiene in Germany. It was the response to several related and widely held beliefs: (1) that humane care for people with disabiltities would enfeeble the "race" because they would survive to pass their disabilities on to their children; (2) that not just mental and physical diseases and so-called defects, but also poverty, criminality, alcoholism, prostitution, and other social problems were based in biology and inherited; and (3) that genetically inferior people were reproducing faster than superior people and would eventually displace them. Although these beliefs were not based in fact, they fueled racist thinking and social programs in Britain and the United States as well as in Germany.

German racial hygiene was founded in 1895, some dozen years after Galton's eugenics, by a physician, Alfred Plötz, and was based on much the same analysis of social problems as the British and American eugenics movements were. In 1924, Plötz started the *Archive of Race- and Sociobiology* (*Archiv für Rassen- und Gesellschaftsbiologie*) and the next year helped found the Society for Racial Hygiene (Gesellschaft für Rassenhygiene). German racial hygiene initially did not concern itself with pre-

venting the admixture of "inferior" races, such as Jews or gypsies, in contrast to the British and American movements where
miscegenation with blacks, Asians, Native Americans, and immigrants of almost any sort was one of the major concerns. The
recommended means for preventing racial degeneration in
Germany, as elsewhere, was sterilization. Around 1930 even
some German socialists and communists supported the eugenic
sterilization of inmates of psychiatric institutions, although the
main impetus came from the Nazis. The active melding of antisemitism and racial hygiene in Germany began during World
War I and accelerated during the 1920s, partly in response to
economic pressures and a scarcity of available positions, which
resulted in severe competition for jobs and incomes among scientists and physicians, many of whom were Jews.

Racial hygiene was established as an academic discipline in
1923, when Fritz Lenz, a physician and geneticist, was appointed
to the newly created Chair of Racial Hygiene at the University
of Munich, a position he kept until 1933, when he moved to the
Chair of Racial Hygiene at the University of Berlin. Lenz, Eugen
Fischer, and Erwin Baer coauthored the most important textbook on genetics and racial hygiene in German. Published in
1921, it was hailed in a review in the *American Journal of Heredity*
in 1928 as "the standard textbook of human genetics" in the
world (quoted in Proctor, 1988, p. 58). In 1931, it was translated into English, and the translation was favorably reviewed
in Britain and the United States despite its blatant racism, or
perhaps because of it. By 1933, eugenics and racial hygiene
were being taught in most medical schools in Germany.

Therefore the academic infrastructure was in place when the
Nazis came to power and began to build a society that gave biologists, anthropologists, and physicians the opportunity to put
their racist and eugenic theories into practice. Looking back on
this period, Eugen Fischer, who directed the Kaiser Wilhelm
Institute for Anthropology, Human Genetics, and Eugenics in
Berlin from 1927 to 1942, wrote in a newspaper article in 1943:
"It is special and rare good luck when research of an intrinsically theoretical nature falls into a time when the general
world view appreciates and welcomes it and, what is more,
when its practical results are immediately accepted as the basis

for governmental procedures" (quoted in Müller-Hill, 1984, p. 64; my translation). It is not true, as has sometimes been claimed, that German scientists were perverted by Nazi racism. Robert Proctor (1988) points out that "it was largely medical scientists who *invented* racial hygiene in the first place" (p. 38; original emphasis).

A eugenic-sterilization law, drafted along the lines of a "Model Sterilization Law" published by Laughlin (the superintendent of Davenport's Eugenics Record Office at Cold Spring Harbor), was being considered in 1932 by the Weimar government. On July 14, 1933, barely six months after Hitler took over, the Nazi government passed its eugenic-sterilization law. This law established genetic health courts (*Erbgesundheitsgerichte*), presided over by a lawyer and two physicians, one of whom was to be an expert on "hereditary pathology" (*Erbpathologie*), whose rulings could be appealed to similar constituted supreme genetic health courts. However, during the entire Nazi period only about 3 percent of lower-court decisions were reversed. The genetic health courts could order the sterilization of people on grounds that they had a "genetically determined" disease, such as "inborn feeble-mindedness, schizophrenia, manic-depressive insanity, hereditary epilepsy, Huntington's disease, hereditary blindness, hereditary deafness, severe physical malformations, and severe alcoholism" (Müller-Hill, 1984, p. 32; my translation). The law was probably written by Dr. Ernst Rüdin, professor of psychiatry and director of the Kaiser Wilhelm Institute for Genealogy and Demography of the German Research Institute for Psychiatry in Munich. The official commentary and interpretation of the law was published under his name and those of an official of the Ministry of the Interior, also a medical doctor, and of a representative of the Health Ministry in the Department of the Interior who was a doctor of laws. All practicing physicians were sent copies of the law and commentaries describing the acceptable procedures for sterilization and castration.

The intent of the law was eugenic, not punitive. Physicians were expected to report patients and their close relatives to the nearest local health court and were fined if they failed to report someone with a so-called hereditary disease. Although some physicians raised the objection that this requirement invaded the doctor-patient relationship, the health authorities argued that this obligation to notify them was no different from requirements that physicans report the incidence of specific infectious diseases or births and deaths. The eugenic measures were to be regarded as health measures pure and simple. And this is the crucial point: the people who designed these policies and the later policies of euthanasia and mass extermination as well as those who oversaw their execution looked on them as sanitary measures, required in this case to cure not individual patients but the collective—the *Volk*—of threats to its health (Lifton, 1987; Proctor, 1988).

As early as 1934, Professor Otmar von Verschuer, then dean of the University of Frankfurt and director of its Institute for Genetics and Racial Hygiene and later the successor of Fischer as director of the Kaiser Wilhelm Institute for Anthropology, Human Genetics, and Eugenics in Berlin, urged that patients should not be looked on, and treated, as individuals. Rather the patient is but "one part of a much larger whole or unity: of his family, his race, his *Volk*" (quoted in Proctor, 1988, p. 105). Minister of the Interior Wilhelm Frisch estimated that at least half a million Germans had genetic diseases, but some experts thought that the true figure was more like one in five, which would be equivalent to thirteen million. In any event, by 1939 some three to four hundred thousand people had been sterilized, with a mortality of about 0.5 percent (Proctor, 1988, pp. 108–109). After that there were few individual sterilizations. Later, large numbers of people were sterilized in the concentration camps, but that was done without benefit of health courts, as part of the program of human experimentation.

The eugenic-sterilization law of 1933 did not provide for sterilization on racial grounds. Nonetheless, in 1937 about five hundred racially mixed children were sterilized; the children had been fathered by black French colonial troops brought to Europe from Africa after World War I to occupy the Rhineland (the so-called *Rheinlandbastarde*).

The first racist eugenic measures were passed in 1935. They were the Nürnberg antimiscegenation, or blood-protection laws, which forbade intermarriage or sexual relations between Jews and non-Jews and forbade Jews from employing non-Jews in their homes. The Nürnberg laws also included a "Law for the Protection of the Genetic Health of the German People," which required premarital medical examinations to detect "racial damage" and required people who were judged "damaged" to marry only others like themselves, provided they first submitted to sterilization. The Nürnberg laws were considered health laws, and physicians were enlisted to enforce them. So-called positive eugenics was practiced by encouraging "genetically healthy" German women to have as many children as possible. They were persuaded to do so by means of propaganda, economic incentives, breeding camps, and strict enforcement of the law forbidding abortion except for eugenic reasons (Koonz, 1987).

The next stage in the campaign of "selection and eradication" was opened at the Nazi party congress in 1935, where plans were made for the "destruction of lives not worth living." The phrase was borrowed from the title of a book published much earlier, in 1920, by Alfred Hoche, professor of psychiatry and director of the Psychiatric Clinic at Freiburg, and Rudolf Binding, professor of jurisprudence at the University of Leipzig. In their book, entitled *The Release for Destruction of Lives Not Worth Living (Die Freigabe zur Vernichtung lebensunwerten Lebens),* these professors argued for killing "worthless" people, whom they defined as those who are "mentally completely dead" and those who constitute "a foreign body in human society" (quoted in Chorover, 1979, p. 97). At the time the program was initiated, the arguments focused on the money wasted in keeping institutionalized (hence "worthless") people alive, for in the early stages the rationale of the euthanasia campaign was economic as much as eugenic. Therefore the extermination campaign was directed primarily at inmates of state psychiatric hospitals and children living in state institutions for the mentally and physically disabled. Jews were specifically excluded because they were not considered worthy of euthenasia. (Here, too, the Nazis were not alone. In 1942, as the last inmates of German mental hospitals were being finished off, Dr. Foster Kennedy, an American psychiatrist writing in the official publication of the American Psychiatric Association, advocated killing retarded children of five and older (Proctor, 1988.) The arguments were phrased in humane terms like these: "Parents who have seen the difficult life of a crippled or feebleminded child must be convinced that though they have the moral obligation to care for the unfortunate creatures, the wider public should not be obliged . . . to assume the enormous costs that long-term institutionalization might entail" (quoted in Proctor, 1988, p. 183). This argument calls to mind the statement by Bentley Glass (1971) which I quoted in the preceding chapter, about parents not having "a right to burden society with a malformed or a mentally incompetent child."

In Germany, the propaganda was subtle and widespread. For example, Proctor (1988, p. 184) cites practice problems in a high school mathematics text published for the school year 1935–36, in which students were asked to calculate the costs to the Reich of maintaining mentally ill people in various kinds of institutions for different lengths of time and to compare the costs of constructing insane asylums and housing units. How is that for relevance?

Although the euthanasia program was planned in the mid-1930s, it was not implemented until 1939, when wartime dislocation and secrecy made it relatively easy to institute such extreme measures. Two weeks before the invasion of Poland an advisory committee commissioned by Hitler issued a secret report recommending that children born with Down syndrome, microcephaly, and various deformities be registered with the Ministry of the Interior. Euthanasia, like sterilization, was to proceed with the trappings of selection. Therefore physicians were asked to fill out questionnaires about all children in their care up to age three who had any of these kinds of disabilities. The completed questionnaires were sent to three-man committees of medical experts charged with marking each form "plus" or "minus." Although none of these "experts" ever saw the children, those whose forms were marked "plus" were transferred to one of a number of institutions where they were killed. Some of the oldest and most respected hospitals in Germany served as such extermination centers. By 1941 the program was expanded to include older children with disabilities and by 1943, to include healthy Jewish children. Also in 1939, evaluation forms were sent to psychiatric institutions for adults for selection and so-called euthanasia.

By September 1941 over seventy thousand inmates had been killed at some of the most distinguished psychiatric hospitals in Germany, which had been equipped for this purpose with gas chambers, disguised as showers, and with crematoria (Lifton, 1986; Proctor, 1988). (When the mass extermination of Jews and other "undesirables" began shortly thereafter, these gas chambers were shipped east and installed at Auschwitz and other extermination camps.) Most patients were gassed or killed by injection with lethal drugs, but a few physicians were reluctant to intervene so actively and let children die of slow starvation and the infectious diseases to which they became susceptible, referring to this as death from "natural" causes. Relatives were notified that their family member had died suddenly of one of a number of infectious diseases and that the body had been cremated for reasons of public health. Nevertheless, rumors began to circulate, and by 1941 hospital killings virtually ceased because of protests, especially from the Church.

There is a direct link between this campaign of "selection and eradication" and the subsequent genocide of Jews, gypsies, communists, homosexuals, and other "undesirables." Early on these people were described as "diseased" and their presence, as an infection or a cancer in the body of the *Volk.* Proctor (1988, p. 194) calls this rationalization "the medicalization of antisemitism." The point is that the Nazi leaders shouted antisemitic and racist propaganda from their platforms, but when it came to devising the measures for ridding the Thousand-Year Reich of Jews, gypsies, and the other undesirables, the task was shouldered by the scientists and physicians who had earlier devised the sterilization and euthanasia programs for the mentally or physically disabled. Therefore, nothing came easier than a medical metaphor: Jews as cancer, Jews as disease. And so the Nazi extermination program was viewed by its perpetrators as a gigantic program in sanitation and public health. It started with quarantining the offending organisms in ghettoes and concentration camps and ended with the extermination of those who did not succumb to the "natural" consequences of the quarantine, such as the various epidemics and hunger.

Yet a measure of selection was practiced throughout the eradication process: It was still *Auslese* as well as *Ausmerze*. At every step choices were made of who could still be used and who had become "worthless." We have read the books and seen the films that show selections being made as the cattle cars emptied the victims into the concentration camps: to work or to die? That is where Joseph Mengele, an M.D./Ph.D., selected the twins and other unfortunates to use as subjects for his scientific experiments at Auschwitz, performed in collaboration with Professor von Verschuer, at that time director of the Kaiser Wilhelm Institute for Anthropology, Human Genetics, and Eugenics in Berlin. And von Verschuer was not the only distinguished scientist who gratefully accepted the human tissues and body fluids provided by Mengele. After the war it became fashionable to characterize the experiments as "bad science," but as Benno Müller-Hill (1984) emphasizes, nothing about them would be considered "bad" were they done with mice. What was "bad" was not their scientific content but the fact that they were being done with "disenfranchised human beings" (p. 97).

Prenatal Testing: Who Should Inhabit the World?

I want to come back to the present, but I needed to go over this history in order to put my misgivings and those of some of the Germans who are opposing genetic testing into the proper perspective. I can phrase the problem best by rephrasing a question Hannah Arendt asks in the epilogue of her commentary on the trial of Adolf Eichmann. Who has the "right to determine who should and who should not inhabit the world?" (1977). That's what it comes down to.

So let me be clear: I am not suggesting that prenatal diagnosis followed by abortion is similar to euthanasia. Fetuses are not people. And a woman must have the right to terminate her pregnancy, whatever her reasons. I am also not drawing an analogy between what the Nazis did and what we and others in

many of the industrialized countries are doing now. Because
the circumstances are different, different things are being done
and for different reasons. But a similar eugenic ideology underlies what happened then and the techniques now being developed. So it is important that we understand how what happened
then came about—and not in some faraway culture that is altogether different from ours but in the heart of Europe, in a
country that has produced artists, writers, composers, philosophers, jurists, scientists, and physicians the equal of any in
the Western world. Given that record, we cannot afford to be
complacent.

Scientists and physicians in this and other countries are once
more engaged in developing the means to decide what lives are
worth living and who should and should not inhabit the world.
Except that now they provide only the tools, while pregnant
women themselves have to make the decisions, euphemistically
called "choices." No one is forced to do anything. A pregnant
woman must merely "choose" whether to terminate a wanted
pregnancy because she has been informed that her future child
will have a disability (although, as I have said before, usually
no one can tell her how severe the disability will be). If she
"chooses" not to take the tests or not to terminate a pregnancy
despite a positive result, she accepts responsibility for whatever
the disability will mean to that child and to her and the rest of
her family. In that case, her child, her family, and the rest
of society can reproach her for having so-to-speak "caused" that
human being's physical pain as well as the social pain he or she
experiences because our society does not look kindly on people
with disabilities.

There is something terribly wrong with this situation, and although it differs in many ways from what went wrong in Germany, at base are similar principles of selection and eradication.
Lest this analogy seem too abstract, let me give a few examples
of how the principle of selection and eradication now works in
practice.

Think of people who have Huntington's disease; as you may
remember they were on the list of people to be sterilized in Germany. Huntington's disease is a degenerative disease of the nervous system and is unusual among hereditary diseases in that it

is inherited as what geneticists call a dominant trait. In other words, even people in whom only one of the pair of genes that is involved with regulating the relevant metabolic processes is affected manifest the disease. Most other gene-mediated diseases, such as Tay-Sachs disease or sickle-cell anemia, are so-called recessives: Only people in whom both members of the relevant pair of genes are affected manifest the disease. In the case of recessive diseases, people with only one affected gene are called carriers: They do not have the disease and usually do not even know that they carry a gene for it. To inherit a recessive disease such as sickle-cell anemia, a child must get an affected gene from each of its parents; to inherit a dominant disease, such as Huntington's disease, it is enough if she or he gets an affected gene from either parent.

The symptoms of Huntington's disease usually do not appear until people are in their thirties, forties, or fifties—in other words, after most people who want to have children have already had one or more. Woody Guthrie had Huntington's disease, but he did not become ill until after he had lived a varied and productive life, produced a large legacy of songs, and fathered his children. At present, there is no cure for Huntington's disease, although scientists have been working to find one. However, a test has been developed that makes it possible to establish with fair reliability whether a person or fetus carries the gene for Huntington's disease, provided a sufficient number of people in that family is willing to be tested.

The existence of this test puts people with a family history of Huntington's disease in an outrageous position: Although they themselves are healthy and do not know whether they will get the disease, they must decide whether to be tested, whether to persuade as many of their relatives as possible to do the same, and whether to test their future child prenatally so they can terminate the pregnancy if the test reveals that the fetus has the gene for Huntington's disease. If it does and they decide on abortion, they are as much as saying that a life lived in the knowledge that one will eventually die of Huntington's disease is not worth living. What does that say about their own life and the lives of their family members who now know that they have the gene for Huntington's disease? If the fetus has the gene and they do not abort, they are knowingly wishing a cruel, degenerative disease on their future child. And if they refuse the test, they can be accused of sticking their heads in the sand. This is an obscene "choice" for anyone to have to make!

Some other inherited diseases also do not become evident until later in life, such as retinitis pigmentosa, a degenerative eye disease. People with this disease are born with normal vision, but their eyesight deteriorates, although usually not until midlife, and they may eventually lose their sight. (People with this disease presumably also were slated for sterilization by the Nazis because it is a form of "hereditary blindness.") There are different patterns of inheritance of retinitis pigmentosa, and prenatal diagnosis is becoming available for one of these patterns and being sought for others. What are prospective parents to do when confronted with the "choice" of aborting a pregnancy because their future child may become blind at some time during its life?

Another, rather different, problem arises with regard to the so-called neural-tube defects (NTDs), a group of developmental disorders which, in fact, are not inherited. They include anencephaly (failure to develop a brain) and spina bifida (failure of the spinal column, and sometimes also the overlying tissues, to close properly). Babies with anencephaly die before birth or shortly thereafter. The severity of the health problems of children who have spina bifida depends on where along the spinal column the defect is located and can vary from life-threatening to relatively mild. The incidence of NTDs varies geographically and tends to be higher in industrialized than in nonindustrialized areas. As we saw in chapter 12, women who carry a fetus with a neural-tube defect have a greater than usual concentration of a specific substance, called alpha-feto-protein, in their blood. A blood test has been developed to detect NTDs prenatally, and California now requires that all pregnant women in the state be offered this test. The women are first counseled about NTDs and about the test and then have to sign a consent or refusal form. If they refuse, that is the end of it. If they consent, they can later refuse to abort the fetus even if the test is positive. This procedure sounds relatively unproblematical, although the requirement to sign a refusal form is coercive. (You cannot walk away; you must say no.) The trouble is that although the test detects virtually all fetuses who have NTDs, it yields a large number of false positive results that suggest that the fetus has a NTD although it does not.

Let us look at some numbers. In California there are about two hundred thousand births a year and the incidence of NTDs is about one per thousand. So, about 200 pregnant women a year carry fetuses with NTDs and 199,800 do not. However, about 5 percent of women test positive on a first test. In other words, if all pregnant women agreed to be tested, 10,000 women would have a positive test, 9,800 of which would be false positives. Those 10,000 women would then have to undergo the stress of worrying as well as further tests in order to determine who among them is in fact carrying a fetus with a NTD. And no test will tell the 200 women whose fetus, in fact, has a NTD how severe their child's health problem will be. All this testing with uncertain results must be offered at this time, when health dollars in California, as elsewhere, have been cut to the bone, and increasing numbers of pregnant women are coming to term with little or no prenatal services of any sort.

The reason I have spelled this problem out in such detail is to make it clear that in many of these situations parents have only the most tenuous basis for making their decisions. Because of the fear of raising a child with a serious disability, many women "choose" to abort a wanted pregnancy if they are told that there is any likelihood whatever that their future child may have a health problem. At times like that we seem to forget that we live in a society in which every day people of all ages are disabled by accidents—at work, on the street, or at home—many of which could be prevented if the necessary money were spent, the necessary precautions taken. What is more, because of the deteriorating economic conditions of poor people and especially women, increasing numbers of babies are born with disabilities that could easily be prevented and are prevented in most other industrialized nations. I question our excessive preoccupation with inherited diseases while callousness and economic mismanagement disable and kill increasing numbers of children and adults.

To say it again, I am not arguing against a woman's right to abortion. Women must have that right because it involves a decision about our bodies and about the way we will spend the rest of our lives. But for scientists to argue that they are developing these tests out of concern for the "quality of life" of future children is like the arguments about "lives not worth living." No one can make that kind of decision about someone else. No one these days openly suggests that certain kinds of people be killed; they just should not be born. Yet that involves a process of selection and a decision about what kinds of people should and should not inhabit the world.

German women, who know the history of Nazi eugenics and how genetic counseling centers functioned during the Nazi period, have organized against the new genetic and reproductive technologies (Duelli Klein, Corea, and Hubbard, 1985). They are suspicious of prenatal testing and counseling centers because some of the scientists and physicians working in them are the same people who designed and implemented the eugenics program during the Nazi period. Others are former co-workers or students of these Nazi professors.

Our history is different, but not different enough. Eugenic thinking is part of our heritage and so are eugenic sterilizations. Here they were not carried over to mass exterminations because we live in a democracy with constitutional safeguards. But, as I mentioned before, even in recent times black, Hispanic, and Native-American women have been sterilized against their wills (Rodriguez-Trias, 1982). We do not exalt the body of the people, as a collective, over that of individuals, but we come dangerously close to doing so when we question the "right" of parents to bear a child who has a disability or when we draw unfavorable comparisons between the costs of care for children with disabilities and the costs of prenatal diagnosis and abortion. We come mighty close when we once again let scientists and physicians make judgments about who should and who should not inhabit the world and applaud them when they develop the technologies that let us implement such judgments. Is it in our interest to have to decide not just whether we want to bear a child but what kind of children to bear? If we try to do that we become entirely dependent on the decisions scientists and physicians make about what technologies to develop and what disabilities to "target." Those decisions are usually made on grounds of professional interest, technical feasibility, and economic and eugenic considerations, not out of a regard for the needs of women and children.

Problems with Selective Abortion

I want to be explicit about how I think a woman's right to abortion fits into this analysis and about some of the connections I see between what the Nazis did and what is happening now. I repeat: A woman must have the right to abort a fetus, whatever her reasons, precisely because it is a decision about her body and about how she will live her life. But decisions about what kind of baby to bear inevitably are bedeviled by overt and unspoken judgments about which lives are "worth living."

Nazi eugenic practices were frankly coercive. The state decided who should not inhabit the world, and lawyers, physicians, and scientists provided the justifications and means to implement these decisions. In today's liberal democracies the situation is different. Eugenic principles are part of our largely unexamined and unspoken preconceptions about who should and who should not inhabit the world, and scientists and physicians provide the ways to put them into practice. Women are expected to implement the society's eugenic prejudices by "choosing" to have the appropriate tests and "electing" not to initiate or to terminate pregnancies if it looks as though the outcome will offend. And to a considerable extent not initiating or terminating these pregnancies may indeed be what women want to do. But one reason we want to is that society promises much grief to parents of children it deems unfit to inhabit the world. People with disabilities, like the rest of us, need opportunities to act in the world, and sometimes that means that they need special provisions and consideration.

So once more, yes, a woman must have the right to terminate a pregnancy, whatever her reasons, but she must also feel empowered not to terminate it, confident that the society will do what it can to enable her and her child to live fulfilling lives. To the extent that prenatal interventions implement social prejudices against people with disabilities they do not expand our reproductive rights. They constrict them.

Focusing the discussion on individualistic questions, such as every woman's right to bear healthy children (which in some people's minds quickly translates into her duty not to "burden society" with unhealthy ones) or the responsibility of scientists and physicians to develop techniques to make that possible, obscures crucial questions such as: How many women have economic access to these kinds of choices? How many have the educational and cultural background to evaluate the information they can get from physicians critically enough to make an informed choice? It also obscures questions about a humane society's responsibilities to satisfy the requirements of people with special needs and to offer them the opportunity to participate as full-fledged members in the culture.

Our present situation connects with the Nazi past in that once again scientists and physicians are making the decisions about what lives to "target" as not worth living by deciding which tests to develop. Yet if people are to have real choices, the decisions that determine the context within which we must choose must not be made in our absence—by professionals, research review panels, or funding organizations. And the situation is not improved by inserting a new group of professionals—bioethicists—between the technical professionals and the public. This public—the women and men who must live in the world that the scientific/medical/industrial complex constructs—must be able to take part in the process by which such decisions are made. Until mechanisms exist that give people a decisive voice in setting the relevant scientific and technical agendas and until scientists and physicians are made accountable to the people whose lives they change, technical innovations do not constitute new choices. They merely replace previous social constraints with new ones.

15

Of Embryos and Women

Until the mid-1970s there was no way scientists could see, much less study, human embryos. After all, fertilization ordinarily happens in the fallopian tubes, deep inside women's bodies. In the usual course of events, no one knows that an embryo has begun to develop until about the time it becomes implanted in the uterus. Scientists got access to embryos only if something went wrong after the embryo had become sufficiently attached for a woman to experience its loss as a miscarriage, and those embryos often had not developed normally in the first place.

All this changed in November 1977, with the birth of Louise Brown, which marked the culmination of more than a decade of research by the British physiologist Robert Edwards, who was joined in the late stages of this work by the obstetrician Patrick Steptoe (Edwards and Steptoe, 1980). The birth of baby Louise showed that in vitro fertilization could be used to circumvent infertility caused by blocked fallopian tubes. With in vitro fertilization came access to human eggs and fertilization, and to human embryos in the first stages of cell division and differentiation. A daunting experience for students of human embryology!

In a moment we shall look at some of the hazards and benefits of the procedure for women, but whatever they may be, there can be no doubt that if there were no medical benefit to in vitro fertilization, scientists could not obtain early human embryos for research. Therefore scientists who want to engage in research with human embryos have a professional interest in stressing the benefits of the in vitro procedure for infertile women or couples. I will begin by discussing the procedure as a fertility measure and then move on to its potential for research and for other kinds of therapies.

The Nature of the Technology

Women do not become candidates for in vitro fertilization until they and their partners have gone through extensive workups to try to determine the reason for their infertility. When the technology was first developed, it was intended specifically for women whose fallopian tubes are blocked (usually because of a previous infection, such as gonorrhea or pelvic inflammatory disease) but whose ovaries and uterus appear to function normally. Because these women ovulate regularly, the idea was that if an egg could be collected before it left the ovary and was fertilized outside the body, the early embryo could then be introduced into the uterus via the cervix and perhaps it would implant and develop as it does ordinarily. Research has concentrated on how to time ovulation so as to know when the eggs are mature and should be collected, what medium is most conducive to fertilization, how long and under what conditions to let the embryo go through its initial cell divisions outside the woman's body, and how best to introduce the embryo into her uterus.

At present, it is usual to stimulate the woman's ovaries by means of hormones, called fertility drugs, to make the ovaries produce more than one egg so that the physician can collect several eggs at one time. The surgical procedure used to collect the eggs is called *laparoscopy* and is done under total anesthesia. The eggs are placed in an appropriate solution, mixed with freshly ejaculated sperm, and incubated at the proper temperature. When some of the eggs have been fertilized and the resulting embryos have begun to divide, a few of them, together with a small amount of the incubation fluid, are sucked into a capillary tube and transferred into the woman's uterus. At present, most clinics transfer three embryos at a time because that has proved to be most effective for achieving a pregnancy, but it also means that more than the usual number of women may end up having twins or, indeed, triplets.

What happens to the embryos that are not implanted? There are several possibilities. Usually, the embryos are considered the property of the couple who has engendered them. The couple can decide to have the clinic freeze and store the embryos that are not implanted for possible use in their future pregnancies. Alternatively, they can donate the extra embryos for use by other couples who cannot produce eggs or sperm. Or they can allow the embryos to be used for research.

Different countries are passing different regulations about the use of embryos. For example, the Swedish government has just ruled that an embryo can be gestated only by the woman who produced the egg and that frozen embryos may not be stored more than a year. It has put off the question of using the embryos for research for a future ruling. In the United States, the use of embryos so far is not regulated, and different clinics go by different rules. For institutions that receive federal funds, research on embryos was supposed to be regulated by an Ethics Advisory Board, convened in 1978. But the board was dissolved when President Reagan came into office and not reconstituted until late in 1988, so there has been no federally funded embryo research. There are no regulations for institutions that are not federally funded.

Over the years, a number of strange ethical dilemmas have hit the news. There was the case of the frozen embryos, stored in Australia, whose progenitors died in a plane crash, leaving a considerable fortune. Were the embryos to be thought of as part of that fortune or were they to be gestated by another woman so they could become its heirs? At present a couple in Tennessee who are seeking a divorce are litigating about which of them can claim as their own seven frozen embryos they engendered together.

Some Health Issues

Because this technique is young, we have no way to assess in what ways and to what extent the various manipulations may affect the long-term health of the children who are produced or of the women who produce them. For one thing the prelimi-

nary workups as well as the eventual pregnancy, if it occurs, require a great deal of monitoring with ultrasound, which may
involve risks, as I pointed out before. Also, it is possible that
stimulation of the ovaries with fertility drugs can increase the
incidence of ovarian cancer. Some people fear that because
the chemical structure of these drugs is closely related to that of
D.E.S., the fetuses these women gestate may later experience
the types of health problems previously associated with the
use of D.E.S. by pregnant women (Bartels, 1988; also see chapter 12).

Scientists do not have a good track record for predicting
problems that can arise from technological interventions in
complicated biological systems, as is evidenced by the fact that
we are in the midst of multiple environmental crises. The physical models are too simplistic and have led to unforeseen problems in the areas of pest control, disposal of organic, chemical,
and nuclear wastes, and health.

In reproductive biology, the nature of the many interacting
processes is poorly understood. Scientists are in no position to
enumerate or describe all the different reactions that must occur at just the right times during the early stages of embryonic
development. But we know that the fertilized egg must begin to
divide into increasing numbers of cells and that these must establish patterns of differentiation that allow the embryo to implant in the uterus and develop into a baby.

The safety of in vitro fertilization and embryo replacement
for humans was not established in animal experiments because
the requirements and details of embryonic development are
different for different animals and for animals and people.
Therefore, the guinea pigs for the in vitro procedure are the
women who provide the eggs, the women who lend their wombs
(who, of course, need not be the same women), and the children who are born.

Some Social Considerations

By present estimates, about ten million Americans are infertile,
or about one in six. In the United States, people are considered
infertile if they have not conceived after a year of regular, un-

protected sexual intercourse. One in five or six couples who meet this definition remain infertile throughout their lives, but for most people the diagnosis is not final. Considerable numbers of women conceive while they go through the preliminary workups for infertility treatments, and some women have become pregnant while on the waiting list for in vitro fertilization (U.S. Congress, Office of Technology Assessment, 1988). A major study published in 1983, showed that over a two- to seven-year period, 41 percent of couples who were undergoing treatments for infertility experienced pregnancies, but so did 35 percent of couples who had decided against treatment (Collins, Wrixon, Lancs, and Wilson, 1983). In France, where a diagnosis of infertility requires two years of regular, unprotected intercourse, as might be expected, the estimated incidence of infertility is considerably lower than in the United States.

Among about a third of infertile couples, the incapacity rests only with the woman, among a third only with the man, and among a third with both. Approximately a third of infertile women have blocked fallopian tubes and when in vitro fertilization was first developed, they were the only ones thought to be candidates for the procedure. This expectation would suggest a limited usefulness for such a complicated and expensive technology. Perhaps for this reason, in recent years the indication for in vitro fertilization has been broadened, and the procedure is now considered appropriate for perfectly healthy women whose partners have a low sperm count and for couples with infertility of unknown origin.

Traditionally, women have been blamed for all instances of infertility, even though the problem as often lies with the man. And women frequently blame themselves, no matter what the reason may be. Now that the in vitro procedure is used for couples where a low sperm count is the problem, there is yet another reason to shift the "blame" to women. Here the doctors got the egg and sperm to join and put the embryo inside the woman, but she cannot make a go of it!

Because of the new range of indications for in vitro fertilization, rather a large proportion of infertile couples have become candidates for this expensive procedure. Yet it has other serious drawbacks, among them, the low success rate. Some of the best clinics claim that 15 to 20 percent of attempts result in babies, but that includes the babies from multiple pregnancies. Some clinics that claim good success rates count so-called chemical pregnancies—that is, early positive pregnancy tests—and do not correct their figures for the high incidence of spontaneous abortions. Some clinics that advertise reasonable success rates in fact cannot claim a single successful birth. The problem is that in the United States in vitro fertilization has become a profitable, entirely unregulated industry. At present, a committee of Congress is conducting a survey of in vitro clinics, and Representative Patricia Schroeder has proposed that the procedure be covered by health insurance. These initiatives may result in some measure of oversight, but we must question the need for this technology altogether.

Advocates of the new technology have begun to argue that every woman has a right to bear a child and that the technology will extend this "right" to a group previously denied it. We have seen in earlier chapters the ambiguities inherent in novel procreative "rights," such as the right to have healthy children. It is not clear to me in what sense we have a "right" to bear children.

There are also questions about the degree to which women have control over in vitro technology. Many women who become candidates for in vitro have had their fallopian tubes damaged by pelvic inflammatory disease, acquired as a result of using a medically prescribed intrauterine device. Having been hurt by one insufficiently tested medical technology, they now seek relief in another, more complicated and even less tested one.

The in vitro technology requires highly skilled professionals and is hard to demystify. There is no way to put control over this technology into the hands of the women who use it. Take an admittedly extreme example: Before Leslie Brown became pregnant with Louise, Drs. Steptoe and Edwards demanded that she not tell anyone about the procedure and made her agree in writing that she would have an abortion if they thought it appropriate. Presumably they did not want to have the entire venture discredited by letting the first baby be born with a disability. In a different vein that also illustrates women's loss of control, the Swedish regulations I mentioned previously stipulate that the procedure may be performed only on married

women or women living in a permanent relationship with a
man, and only by using the sperm of their partner. No matter
whether one approves of the societal interests that these regulations are intended to preserve, which I do not, they clearly have
no relevance to the technical content of the procedure. They
merely extend into new areas the state's power to regulate
women's procreative lives.

The pregnancies that result from in vitro fertilization are
considered "precious," and, therefore, a disproportionate
number of them end in Caesareans because obstetricians feel in
better control of births they can initiate and carry through than
of those that develop according to women's internal rhythms.
I have trouble interpreting this degree of subservience to medical and societal regulation as an increase in procreative freedom or choice.

Embryo Research

Embryos that have been produced in vitro but are not implanted constitute the only available source of human embryos
for research. At present, a number of European parliaments
are debating legislation to outlaw research on human embryos.
Several others are considering, or have passed, laws that permit
research on embryos up to fourteen days after fertilization. As
we saw earlier, in the United States, because the charter of the
Ethics Advisory Board lapsed at the beginning of President
Reagan's term in office, human embryo research has been effectively forbidden in institutions that receive federal funds but is
unregulated in those that do not (Krimsky, Hubbard, and
Gracey, 1989).

The issues are not easy to sort out. People who think that human life begins at conception tend to argue that no one—egg
and sperm donors, scientists, or medical people—can give permission to use a human embryo for research. I do not share this
opinion but think the embryo's progenitors have that right.
However, I worry about the potential for exploitation if medical scientists are allowed to persuade people, or even pay them,
to produce embryos for research.

Some people have another worry. If embryologists learn how to get early embryos to develop outside the womb, while neo-natologists learn how to keep less and less developed, premature babies alive, how long before we can expect the two ends to meet? Is this the road to designing plastic wombs and dispensing with human gestation?

I do not think it is. True, scientists need to understand the two ends better before they can explore the middle, but it is much easier to sustain the beginning and end of gestation outside the womb than the middle. Scientists are learning how to keep early embryos alive before the rudiments of the main organ systems are laid down. At the other end, they are learning how to enable premature babies (who until recently would have been miscarried, dying fetuses) to live outside their mother's womb. Neither effort approaches the all-important question of how the organ systems begin to differentiate and develop, and I do not think scientists are nearly ready to tackle that one. Therefore I do not think that learning about the early stages of human embryonic development raises this particular threat. Nor is the threat of exploitation new. We will have to guard against it in this situation, as elsewhere.

The manipulation of early human embryos for purposes of so-called gene therapy, however, does raise new issues. It could open up the "not so sci-fi fantasy" I laid out in chapter 13. Conception could be routinely moved out of women's bodies, so that early embryos could be submitted to a battery of genetic tests. The number and availability of these tests is bound to increase rapidly as the Human Genome Initiative gets under way. At that point any embryo found to have a genetic "defect" could either be discarded or be "treated" by introducing the "normal" gene. I have put all these words in quotes to emphasize the normative, eugenic ideology that underlies the entire effort.

The British embryologist Anne McLaren (1987) argues that it will never make sense to repair genetic "defects" in early embryos, that it will always be better to discard such embryos and only implant "normal" ones. But there is no question that some scientists will want to insert genes into early embryos if only to learn how to do it most effectively.

Scientists differentiate between two kinds of so-called gene

therapy. There is somatic gene therapy, in which one introduces a gene into a specific tissue, say the bone marrow or
muscle of an adult or a child, in an attempt to repair a genetic
lesion in that tissue. And then there is germ-line gene therapy,
in which genes would be inserted into a fertilized egg or an
early embryo. At present, neither has been done successfully,
but there is a significant difference between them. If somatic
therapy can be made to work, it will affect only the individual
who gets the gene. In germ-line therapy, however, the new
gene would enter all, or most, cells of the person into whom the
embryo will develop and will be passed on to her or his offspring.

A few scientists are trying to develop somatic gene therapy.
Most scientists disavow any intention to do germ-line therapy.
Yet in 1986, when the Committee for Responsible Genetics, a
public-interest group in the United States, asked the Recombinant DNA Advisory Committee of the National Institutes of
Health, which is the organization charged with regulating genetic engineering of all kinds, to state that it will not accept any
proposals to do germ-line gene therapy, the Advisory Committee refused (*Federal Register*, 1986).

There needs to be a broadly based debate about whether experiments with human embryos are ethically admissible, and if
they are, what kinds should be allowed. I am not overly concerned about the dangers to the embryos of experiments done
for the purpose of learning about the nature of early development, provided the embryos are not implanted or allowed to
survive beyond fourteen days and provided their progenitors
freely give their consent.

I am concerned, however, about experiments that involve the
use of embryonic tissue for therapeutic or cosmetic purposes.
Some scientists think that embryonic tissue may be suitable for
repairing metabolic diseases or undesired manifestations of
aging. Here again, the issue for me is not the use of the tissue
per se. My concerns focus on how much control the progenitors
will have, how expensive the procedures will be, and how to
avoid crass commercialization of human embryos. Just as we
have so far avoided selling human kidneys or hearts for trans-
plants, it is important that we not begin to sell embryos and
fetuses.

All these questions suggest that it is unrealistic to try to evaluate the societal costs and benefits of in vitro fertilization without also looking at the directions in which the procedure is likely to be taken. We are not just talking about babies for childless couples. We are talking about a range of new issues in medical science and practice that must be evaluated by social and political mechanisms we do not have but urgently need.

The biology of procreation is embedded in human relationships and cultural realities. It is not just a question of hormones, eggs, sperm, fallopian tubes, uteri. . . . The ground under procreation is shifting, and the terms are being renegotiated. Women must be part of this process. There are several sides to these issues, and women do not speak with one voice. But that is all the more reason why we must make ourselves heard, lest we slip into a brave new world in whose creation we have no part.

Some Final Thoughts

I want to emphasize once more that biology is profoundly political. Biologists have the authority to tell us what is natural and what is human. They sort nature from culture, and what is more political than that? Biologists have been able to wrap the mantle of science around racism and sexism by inventing significant characteristics to describe and sort different groups of people, and have performed the measurements that made the answers come out the way political prejudices predicted they would.

Yet it would be foolish to deny that science and technology have the potential to improve our lives. The problem is not that scientists practice their skills but that they have been allowed to constitute a priesthood. Scientists are thought to be immune from ideological and political influences because the scientific methodology is supposed to neutralize their ordinary human commitments. Scientific consensus and peer review are expected to eliminate bias, but these devices can guard only against personal or idiosyncratic quirks, not against beliefs shared in the scientific community as well as by powerful segments of the rest of society.

Scientists tend to label as antiscience anyone who criticizes the beliefs, methods, or results accepted by the scientific elite. By drawing sharp lines, they can discredit even highly accredited scientists. Needless to say, this discreditation does not work every time, but it tends to create confusion, which makes it difficult, if not impossible, for nonscientists to know whom to believe. Once confused, the public can be readily persuaded that it had best leave all the important decisions to the experts.

Scientists are an interest group whose primary concern is to continue their work and perpetuate their kind. If challenged, they will set a few moments aside to tell the public how important and useful their work is. For this purpose almost any story will do, just so long as it is likely to allay people's concerns. A striking example is described in a report in *Science* magazine about a political battle between German embryologists and geneticists on one side and the West German parliament on the other (Kirk, 1988). The Bundestag is considering legislation that would outlaw research on human embryos. As might be expected, German jurists and politicians feel sensitive about the history of human experimentation in Germany, and all major political parties favor the legislation. The scientists, however, unselfconsciously talk about the need for German science to remain competitive. In the article, the director of the Cologne Institute of Genetics is quoted as saying: "Embryology and genetics have a bad reputation in West Germany. It's not clear to many people just how well the self-control mechanism of science works." This from a German scientist, apparently oblivious to how desperately the "self-control mechanism" failed when scientists agreed among themselves and with their government that painful and deadly human experimentation was in the interests of science and the Reich.

We need not go abroad for examples of the failure of scientists to overcome widely held prejudices. In Tuskegee, Alabama, four hundred black men with syphilis, all of them poor and most of them illiterate, were observed from 1932 until 1972 without being told that they had the disease or being treated for it although it was curable. This infamous experiment was carried out by the U.S. Public Health Service, at times in collaboration with various other federal, state, and local health agencies, in order to document the progress of syphilis all the way to autopsy (Jones, 1981). In San Antonio, Texas, in the 1960s, Dr. Joseph Goldzieher of the Southwest Foundation for Research and Education gave placebos (dummy pills) to a group of seventy-six poor, Hispanic women who came to his clinic for birth-control pills and did not tell them they were part of an experiment. He merely suggested that they use vaginal cream for added protection and did nothing to help the eleven women who got pregnant as a result. Yet the U.S. Agency for International Development continued to fund his research (Seaman and Seaman, 1978).

Nor can we take comfort in the thought that this is past history. On April 3, 1989, *The Nation* magazine carried a story about an experiment being conducted in Olongapo City near the Subic Naval Base in the Philippines. According to this report, eight women, apparently former prostitutes, who test positive for the human immunodeficiency virus (HIV) that is responsible for AIDS are being tested every three months for three days at the U.S. Naval Hospital in Manila. "They are given no word on what their diagnosis means; they are told that they are recovering and are given job-training classes and urged to plan for the future" so as to get them to "think positively" (Sturdevant, 1989).

My reason for telling these stories is to point out once again that scientific objectivity offers no protection against prejudices scientists share with their society. Perhaps we should not even look at such extreme examples but simply remember that scientists routinely have invented criteria and made measurements that underwrite racial and gender inequalities (Gould, 1981; Hubbard and Lowe, 1979).

For science and technology to be useful and responsive to people's needs, scientists, along with everyone else, will have to recognize that science is no more immune from ideological commitments than are other human activities and that we therefore need better and more democratic mechanisms than we now have to decide what science needs to be done and how best to do it.

Bibliography

Arendt, Hannah. 1977. *Eichmann in Jerusalem: A Report on the Banality of Evil.* New York: Penguin.

Asch, Adrienne. 1988. "Reproductive Technology and Disability." In Sherrill Cohen and Nadine Taub, eds., *Reproductive Laws for the 1990s.* Clifton, N. J.: Humana Press.

Asch, Adrienne, and Michelle Fine. 1988. "Introduction: Beyond Pedestals." In Michelle Fine and Adrienne Asch, eds., *Women with Disabilities.* Philadelphia: Temple University Press.

Atwood, Margaret. 1986. *The Handmaid's Tale.* Boston: Houghton Mifflin.

Banta, David. 1981. "Benefits and Risks of Electronic Fetal Monitoring." In Helen B. Holmes, Betty B. Hoskins, and Michael Gross, eds., *The Custom-Made Child?* Clifton, N. J.: Humana Press.

Barash, David. 1979. *The Whispering Within.* New York: Harper & Row.

Bartels, Ditta. 1988. "Built-In Obsolescence: Women, Embryo Production, and Genetic Engineering." *Reproductive and Genetic Engineering* 1:141–152.

Benston, Margaret Lowe. 1986. "Questioning Authority: Feminism and Scientific Experts." *Resources for Feminist Research* 15(3):71–73.

Birke, Lynda. 1986. *Women, Feminism, and Biology.* New York: Methuen.

Blatt, Robin J. R. 1988. *Prenatal Tests.* New York: Random House, Vintage Books.

Bleier, Ruth. 1984. *Science and Gender.* New York: Pergamon Press.

Blumberg, Bruce D., Mitchell S. Golbus, and Karl H. Hansen. 1975. "The Psychological Sequelae of Abortion Performed for a Genetic Indication." *American Journal of Obstetrics and Gynecology* 122:799–808.

Bolsen, Barbara. 1982. "Question of Risk Still Hovers over Routine Prenatal Use of Ultrasound." *Journal of the American Medical Association* 247: 2195–2197.

Boston Globe. 1988. "Lead Levels Called Threat to Fetuses." July 19:73.

Boston Women's Healthbook Collective. 1984. *The New Our Bodies, Ourselves.* New York: Simon & Schuster.

Bowes, Watson A., Jr., and Brad Selegstad. 1981. "Fetal versus Maternal Rights: Medical and Legal Perspectives." *Obstetrics and Gynecology* 58:209–214.

British Medical Association. 1886. "Fifty-Fourth Annual Meeting." *The Lancet,* August 14: 314–315.

Cannon, Walter B. 1939. *The Wisdom of the Body.* New York: Norton.

Children's Defense Fund. 1989. *A Vision for America's Future.* Washington,

D.C.: Children's Defense Fund.

Chodorow, Nancy. 1978. *The Reproduction of Mothering.* Berkeley: University of California Press.

Chorover, Stephan L. 1979. *From Genesis to Genocide.* Cambridge, Mass.: MIT Press.

Clarke, Edward H. 1874. *Sex in Education; or, A Fair Chance for the Girls.* Boston: James R. Osgood and Co.

Collins, John A., William Wrixon, Lynn B. Lanes, and Elaine H. Wilson. 1983. "Treatment-Independent Pregnancy Among Infertile Couples." *New England Journal of Medicine* 309:1201–1206.

Commoner, Barry. 1968. "Failure of the Watson-Crick Theory as a Chemical Explanation of Inheritance." *Nature* 220:334–340.

Crick, F. H. C., and J. D. Watson. 1954. "The Complementary Structure of Deoxyribonucleic Acid." *Proceedings of the Royal Society* A223:80–96.

Darwin, Charles. n.d. *The Origin of Species [1859] and the Descent of Man [1871].* New York: Modern Library Edition.

Darwin, Francis, ed. 1958. *The Autobiography of Charles Darwin.* New York: Dover.

Davis, Angela Y. 1981. *Women, Race, and Class.* New York: Random House.

Dawkins, Richard. 1976. *The Selfish Gene.* Oxford: Oxford University Press.

de Beauvoir, Simone. 1953. *The Second Sex.* New York: Knopf.

DeLee, Joseph. 1920. "The Prophylactic Forceps Operation." *American Journal of Obstetrics and Gynecology* 1:34–44.

d'Emilio, John, and Estelle B. Freedman. 1988. *Intimate Matters: A History of Sexuality in America.* New York: Harper & Row.

Dinnerstein, Dorothy. 1976. *The Mermaid and the Minotaur.* New York: Harper & Row.

Duelli Klein, Renate, Gena Corea, and Ruth Hubbard. 1985. "German Women say No to Gene and Reproductive Technology: Reflections on a Conference in Bonn, West Germany, April 19–21, 1985." *Feminist Forum: Women's Studies International Forum* 8(3):I–IV.

Edwards, Robert, and Patrick Steptoe. 1980. *A Matter of Life: The Sensational Story of the World's First Test-Tube Baby.* London: Sphere Books.

Ehrenreich, Barbara, and Deirdre English. 1978. *For Her Own Good.* Garden City, N.Y.: Doubleday, Anchor Books.

Eiseley, Loren. 1961. *Darwin's Century.* Garden City, N.Y.: Doubleday.

Fausto-Sterling, Anne. 1985. *Myths of Gender.* New York: Basic Books.

Federal Register. 1986. 51(122):23210–23211.

Firestone, Shulamith. 1972. *The Dialectic of Sex.* New York: Morrow.

Fleck, Ludwik. [1935] 1979. *Genesis and Development of a Scientific Fact.* Chicago: University of Chicago Press.

Fletcher, Joseph F. 1974. *The Ethics of Genetic Control.* Garden City, N.Y.: Doubleday.

———. 1980. "Knowledge, Risk, and the Right to Reproduce: A Limiting Principle." In Aubrey Milunsky and George J. Annas, eds., *Genetics and the Law II.* New York: Plenum.

Foucault, Michel. 1978. *The History of Sexuality.* New York: Random House, Vintage Books.

Franklin, R. E., and R. G. Gosling. 1953. "Molecular Configuration in Sodium Thymonucleate." *Nature* 171:740–741.

Freire, Paulo. 1985. *The Politics of Education.* South Hadley, Mass.: Bergin and Garvey.

Fried, Barbara. 1979. "Boys Will Be Boys Will Be Boys: The Language of Sex and Gender." In Ruth Hubbard, Mary Sue Henifin, and Barbara Fried, eds. *Women Look at Biology Looking at Women.* Cambridge, Mass.: Schenkman.

Frisch, Rose. 1988. "Body Fat, Menarche, Fitness and Fertility." *Human Reproduction* 2:521–533.

Gallagher, Janet. 1987. "Prenatal Invasions and Interventions: What's Wrong with Fetal Rights." *Harvard Women's Law Journal* 10:9–58.

———. 1988. "Position Paper: Fetus as Patient." In Sherrill Cohen and Nadine Taub, eds., *Reproductive Laws for the 1990s.* Clifton, N.J.: Humana Press.

Galton, Francis. 1883. *Inquiries into Human Faculty.* London: Macmillan.

Geist, Valerius. 1971. *Mountain Sheep.* Chicago: University of Chicago Press.

Glass, Bentley. 1971. "Science: Endless Horizons or Golden Age?" *Science* 171:23–29.

Goleman, Daniel. 1984. "Schizophrenia: Early Signs Found." *New York Times,* December 11:C1.

Gombrich, Ernst H. 1963. *Meditations on a Hobby Horse.* London: Phaidon Press.

Gordon, Linda. 1976. *Woman's Body, Woman's Right.* New York: Grossman.

Gould, Stephen J. 1981. *The Mismeasure of Man.* New York: Norton.

Goy, Robert W., and Bruce S. McEwen. 1980. *Sexual Differentiation of the Brain.* Cambridge: MIT Press.

Graham, Loren R. 1978. "Concerns about Science and Attempts to Regulate Inquiry." *Daedalus* 107(2): 1–21.

Gwaltney, John Langston 1980. *Drylongso: A Self-Portrait of Black America.* New York: Random House, Vintage Books.

Haldane, J. B. S. 1942. *New Paths in Genetics.* New York: Harper and Brothers.

Haraway, Donna. 1979. "The Biological Enterprise: Sex, Mind, and Profit from Human Engineering to Sociobiology." *Radical History Review,* Spring/Summer: 206–237.

Harding, Sandra. 1986. *The Science Question in Feminism.* Ithaca, N.Y.: Cornell University Press.

Harrison, Michael R. 1982. "Unborn: Historical Perspective of the Fetus as Patient." *Pharos,* Winter: 19–24.

Harrison, Michael R., Mitchell S. Golbus, and Roy A. Filly. 1981. "Management of the Fetus with a Correctable Congenital Defect." *Journal of the American Medical Association* 246:744–747.

Harrison, Michael R., et al. 1982. "Fetal Treatment 1982." *New England Journal of Medicine* 307:1651–1652.

Henifin, Mary Sue. 1986. "Wrongful Life Cases and the Courts." *Women and Health* 11:97–102.

Henifin, Mary Sue, Ruth Hubbard, and Judy Norsigian. 1988. "Position Paper: Prenatal Screening." In Sherrill Cohen and Nadine Taub, eds., *Reproductive Laws for the 1990s*. Clifton, N. J.: Humana Press.

Herrnstein, Richard J. 1971. "I.Q." *Atlantic Monthly* 228(3) (September): 43–64.

Herschberger, Ruth. 1948. *Adam's Rib*. New York: Harper & Row.

Hofstadter, Richard. 1955. *Social Darwinism in American Thought*. Boston: Beacon Press.

Hollingworth, Leta S. 1916. "Social Devices for Impelling Women to Bear and Rear Children." *American Journal of Sociology* 22:19–29.

Howell, Nancy. 1979. *Demography of the Dobe !Kung*. New York: Academic Press.

Howells, William. 1973. *Evolution of the Genus* Homo. Reading, Mass.: Addison-Wesley.

Hrdy, Sarah Blaffer. 1981. *The Woman That Never Evolved*. Cambridge: Harvard University Press.

———. 1986. "Empathy, Polyandry, and the Myth of the Coy Female." In Ruth Bleier, ed., *Feminist Approaches to Science*. New York: Pergamon Press.

Hubbard, Ruth. 1976. "*Rosalind Franklin and DNA* by Anne Sayre." *Signs* 2:229–237.

———. 1979. "Have Only Men Evolved?" In Ruth Hubbard, Mary Sue Henifin, and Barbara Fried, eds., *Women Look at Biology Looking at Women*. Cambridge, Mass.: Schenkman.

Hubbard, Ruth, and Marian Lowe, eds. 1979. *Genes and Gender II: Pitfalls in Research on Sex and Gender*. Staten Island, N.Y.: Gordian Press.

Huxley, Julian. 1941. "The Vital Importance of Eugenics." *Harper's Monthly* 163(August):324–331.

Irvine, William. 1972. *Apes, Angels, and Victorians*. New York: McGraw-Hill.

Jackson, Stevi. 1982. *Childhood and Sexuality*. Oxford: Blackwell.

Jensen, Arthur R. 1969. "How Much Can We Boost IQ and Scholastic Achievement?" *Harvard Educational Review* 39 (Winter):1–123.

Jones, James H. 1981. *Bad Blood*. New York: Free Press.

Keller, Evelyn Fox. 1985. *Reflections on Gender and Science*. New Haven, Conn.: Yale University Press.

Kenen, Regina. 1981. "A Look at Prenatal Diagnosis within the Context of Changing Parental and Reproductive Norms." In Helen B. Holmes, Betty B. Hoskins, and Michael Gross, eds., *The Custom-Made Child?* Clifton, N.J.: Humana Press.

Kevles, Daniel J. 1985. *In the Name of Eugenics: Genetics and the Uses of Human Heredity*. New York: Knopf.

Kinsey, Alfred C., Wardell B. Pomeroy, and Clyde E. Martin. 1948. *Sexual Behavior in the American Male*. Philadelphia: Saunders.

Kinsey, Alfred C., Wardell B. Pomeroy, Clyde E. Martin, and Paul H.

Gebhard. 1953. *Sexual Behavior in the American Female.* Philadelphia: Saunders.

Kirk, Don. 1988. "West Germany Moving to Make IVF Research a Crime." *Science* 241:406.

Kolder, Veronika E. B., Janet Gallagher, and Michael T. Parsons. 1987. "Court-Ordered Obstetrical Interventions." *New England Journal of Medicine* 316p. 1192–1196.

Konner, Melvin, and Carol Worthman. 1980. "Nursing Frequency, Gonadal Function, and Birth Spacing among !Kung Hunter-Gatherers." *Science* 207:788–790.

Koonz, Claudia. 1987. *Mothers in the Fatherland: Women, the Family and Nazi Politics.* New York: St. Martin's Press.

Krimsky, Sheldon, Ruth Hubbard, and Colin Gracey. 1989. "Fetal Research in the United States: A Historical and Ethical Perspective." *Genewatch* 5(4–5):1–3, 8–10.

Labson, Lucy H. 1983. "Today's View in Maternal and Fetal Medicine." *Patient Care* 15 (January):105–121.

Lancaster, Jane B. 1975. *Primate Behavior and the Emergence of Human Culture.* New York: Holt, Rinehart & Winston.

Lane, Ann J., ed. 1977. *Mary Ritter Beard: A Sourcebook.* New York: Schocken.

Leacock, Eleanor Burke, ed. 1972. *The Origin of the Family, Private Property, and the State* [1884], by Frederick Engels. New York: International Publishers.

Leavitt, Judith Walzer. 1987. *Brought to Bed: Childbearing in America, 1750–1950.* New York: Oxford University Press.

Legal Advisors Committee, Concern for Dying. 1983. "The Right to Refuse Treatment: A Model Act." *American Journal of Public Health* 73:918–921.

LeGuin, Ursula K. 1968. *A Wizard of Earthsea.* Emeryville, Calif.: Parnassus Press.

Lewontin, Richard C. 1974. "The Analysis of Variance and the Analysis of Causes." *American Journal of Human Genetics* 26:400–411.

Lewontin, Richard C., Steven Rose, and Leon J. Kamin. 1984. *Not in Our Genes.* New York: Pantheon.

Lieberman, J. R., M. Mazor, W. Chaim, and A. Cohen. 1979. "The Fetal Right to Live." *Obstetrics and Gynecology* 53:515–517.

Lifton, Robert J. 1986. *The Nazi Doctors.* New York: Basic Books.

Lowe, Marian. 1982. "Social Bodies: The Interaction of Culture and Women's Biology." In Ruth Hubbard, Mary Sue Henifin, and Barbara Fried, eds., *Biological Woman—The Convenient Myth.* Cambridge, Mass.: Schenkman.

Lowe, Marian, and Ruth Hubbard. 1979. "Sociobiology and Biosociology: Can Science Prove the Biological Basis of Sex Differences in Behavior?" In Ruth Hubbard and Marian Lowe, eds., *Genes and Gender II: Pitfalls in Research on Sex and Gender.* Staten Island, N.Y.: Gordian Press.

———. 1983. *Woman's Nature: Rationalizations of Inequality.* New York: Pergamon Press.

Ludmerer, Kenneth M. 1972. *Genetics and American Society.* Baltimore: Johns Hopkins University Press.

McDonald, Marian, ed. 1981. *For Ourselves, Our Families, and Our Future.* Boston: Red Sun Press.

McLaren, Anne. 1987. "Can We Diagnose Genetic Disease in Pre-Embryos?" *New Scientist* 116(December 10):42–47.

Martin, Emily. 1987. *The Woman in the Body.* Boston: Beacon Press.

Mead, Margaret. 1949. *Male and Female.* New York: Dell.

Mendel, Agata. 1980. *Les Manipulations Génétiques.* Paris: Editions du Seuil.

Mendel, Gregor. [1865] 1950. *Experiments in Plant Hybridisation.* Translation prepared by the Royal Historical Society of London, with notes by W. Bateson. Cambridge: Harvard University Press.

Messer-Davidow, Ellen. 1987. "The Philosophical Bases of Feminist Literary Criticism." *New Literary History* 19:63–103.

Miller, Alice. 1984. *For Your Own Good.* New York: Farrar, Straus & Giroux.

———. 1986. *Thou Shalt Not Be Aware.* New York: Meridian.

Money, John, and Anke A. Ehrdardt. 1972. *Man & Woman, Boy & Girl.* Baltimore: Johns Hopkins University Press.

Morgan, Elaine. 1973. *The Descent of Woman.* New York: Bantam.

Morgan, Thomas Hunt. 1909. "What Are 'Factors' in Mendelian Explanations?" *American Breeders Association* 5:365.

———. 1926. *The Theory of the Gene.* New Haven, Conn: Yale University Press.

Morgan, Thomas Hunt, A. H. Sturtevant, H. J. Muller, and C. B. Bridges. 1915. *The Mechanism of Mendelian Heredity.* New York: Henry Holt and Co.

Morison, Robert S. 1978. "Introduction." *Daedalus* 107(2):vii–xvi.

Muller, H. J. [1913]. 1962. "Principles of Heredity." In H. J. Muller, ed., *Studies in Genetics.* Bloomington: University of Indiana Press.

Müller-Hill, Benno. 1984. *Tödliche Wissenschaft.* Reinbek, West Germany: Rowohlt. (Translation 1988. *Murderous Science.* Oxford: Oxford University Press.)

Needham, Joseph. 1969. "Human Law and the Laws of Nature." *The Grand Titration.* London: Allen and Unwin.

Neilsen, Caroline C. 1981. "An Encounter with Modern Medical Technology: Women's Experiences with Amniocentesis." *Women and Health* 6:109–124.

Nelson, Harry. 1983. "Some Schizophrenia Linked to Prenatal Changes in Brain Cells." *Boston Globe,* June 7:8.

Newman, Louise M., ed. 1985. *Men's Ideas/Women's Realities: Popular Science, 1870–1915.* New York: Pergamon Press.

Nsiah-Jefferson, Laurie. 1988. "Reproductive Laws, Women of Color, and Low-Income Women." In Sherrill Cohen and Nadine Taub, eds., *Reproductive Laws for the 1990s.* Clifton, N.J.: Humana Press.

Oakley, Kenneth P. 1972. *Man the Toolmaker.* London: British Museum.

Olby, Robert C. 1966. *Origins of Mendelism.* London: Constable.

———. 1974. *The Path to the Double Helix.* London: Macmillan.

Pauling, Linus. 1970. "Fifty Years of Progress in Structural Chemistry and Molecular Biology." *Daedalus* 99:988–1014.

Perry, Ruth. 1979. "The Veil of Chastity: Mary Astell's Feminism." *Studies in Eighteenth-Century Culture* 9:25–45.

Petchesky, Rosalind Pollack. 1987. "Foetal Images: The Power of Visual Culture in the Politics of Reproduction." In Michelle Stanworth, ed., *Reproductive Technologies*. Minneapolis: University of Minnesota Press.

Proctor, Robert N. 1988. *Racial Hygiene: Medicine under the Nazis.* Cambridge: Harvard University Press.

Rapp, Rayna. 1984. "XYLO: A True Story." In Rita Arditti, Renate Duelli Klein, and Shelley Minden, eds., *Test-Tube Women*. London: Pandora Press.

Reed, Charles B. 1920. "The Induction of Labor at Term." *American Journal of Obstetrics and Gynecology* 1:24–33.

Rhoads, George G., et al. 1989. "The Safety and Efficacy of Chorionic Villus Sampling for Early Prenatal Diagnosis of Cytogenetic Abnormalities." *New England Journal of Medicine* 320:609–617.

Rich, Adrienne. 1976. *Of Woman Born.* New York: Norton.

Robertson, John A. 1983. "Procreative Liberty and the Control of Conception, Pregnancy, and Childbirth." *Virginia Law Review* 69:405–464.

Rodriguez-Trias, Helen. 1982. "Sterilization Abuse." In Ruth Hubbard, Mary Sue Henifin, and Barbara Fried, eds., *Biological Woman—The Convenient Myth*. Cambridge, Mass.: Schenkman.

Rossiter, Margaret W. 1982. *Women Scientists in America: Struggles and Strategies to 1940.* Baltimore: Johns Hopkins University Press.

Rothman, Barbara Katz. 1982. *In Labor: Women and Power in the Birthplace.* New York: Norton.

———. 1984. "The Meanings of Choice in Reproductive Technology." In Rita Arditti, Renate Duelli Klein, and Shelley Minden, eds., *Test-Tube Women*. London: Pandora Press.

———. 1986.*TheTentative Pregnancy: Prenatal Diagnosis and the Future of Motherhood*. New York: Norton.

———. 1989. *Recreating Motherhood: Ideology and Technology in a Patriarchal Society.* New York: Norton.

Sahlins, Marshall. 1976. *The Use and Abuse of Biology.* Ann Arbor: University of Michigan Press.

Sayre, Anne. 1975. *Rosalind Franklin and DNA.* New York: Norton.

Schrödinger, Erwin. 1944. *What Is Life?* London: Cambridge University Press.

Science. 1981. 211 (4488, March 20):1263–1324.

Seaman, Barbara, and Gideon Seaman. 1978. *Women and the Crisis in Sex Hormones.* New York: Bantam.

Shaw, Margery W. 1980. "The Potential Plaintiff: Preconception and Prenatal Torts." In Aubrey Milunsky and George J. Annas, eds., *Genetics and the Law II*. New York: Plenum.

Shostak, Marjorie. 1981. *Nisa.* Cambridge: Harvard University Press.

Shriner, Thomas L. 1979. "Maternal Rights versus Fetal Rights—A Clinical Dilemma." *Obstetrics and Gynecology* 53:518–519.

Sinsheimer, Robert L. 1978. "The Presumptions of Science." *Daedalus* 107(2):23–35.

Sjöström, Henning, and Robert Nilsson. 1972. *Thalidomide and the Power of the Drug Companies.* Hammondsworth, England: Penguin.

Smuts, Barbara B. 1985. *Sex and Friendship in Baboons.* New York: Aldine.

Stadler, L. J. 1954. "The Gene." *Science* 120:811–819.

Stellman, Jeanne M., and Mary Sue Henifin. 1982. "No Fertile Women Need Apply: Employment Discrimination and Reproductive Hazards in the Workplace." In Ruth Hubbard, Mary Sue Henifin, and Barbara Fried, eds., *Biological Woman—The Convenient Myth.* Cambridge, Mass.: Schenkman.

Sturdevant, Saundra. 1989. "The Bar Girls of Subic Bay." *The Nation* 248(April 3):444–446.

Tanner, Nancy M. 1981. *On Becoming Human.* New York: Cambridge University Press.

Tanner, Nancy M., and Adrienne L. Zihlman. 1976. "Women in Evolution. 1. Innovation and Selection in Human Origins." *Signs* 1:585–608.

Terman, Lewis M. 1924. "The Conservation of Talent." *School and Society* 19(483):359–364.

U.S. Congress, Office of Technology Assessment. 1988. *Infertility: Medical and Social Choices.* Washington, D.C.: U.S. Government Printing Office.

Walsh, Mary Roth. 1977. *Doctors Wanted: No Women Need Apply.* New Haven, Conn.: Yale University Press.

———. 1979. "The Quirls of a Woman's Brain." In Ruth Hubbard, Mary Sue Henifin, and Barbara Fried, eds., *Women Look at Biology Looking at Women.* Cambridge, Mass.: Schenkman.

Watson, James D. 1966. "Growing Up in the Phage Group." In J. Cairns, G. S. Stent, and J. D. Watson, eds., *Phage and the Origins of Molecular Biology.* Cold Spring Harbor, N.Y.: Cold Spring Harbor Laboratory of Quantitative Biology.

———. 1968. *The Double Helix.* New York: Atheneum.

———. 1989. Quoted in Pamela Zurer, "Panel Plots Strategy for Human Genome Studies." *Chemical and Engineering News,* January 9:5.

Watson, James D., and F. H. C. Crick. 1953. "A Structure for Deoxyribose Nucleic Acid." *Nature* 171:737—738.

Watts, Alan W. [1958] 1970. *Nature, Man and Woman.* New York: Pantheon, Vintage.

Weeks, Jeffrey. 1977. *Coming Out: Homosexual Politics in Britain from the Nineteenth Century to the Present.* London: Quartet Books.

———. 1981. *Sex, Politics, and Society: The Regulation of Sexuality Since 1900.* New York: Longman.

Weismann, August. 1893. *The Germ Plasm.* New York: Charles Scribner's Sons.

Wertz, Richard W., and Dorothy C. Wertz. 1977. *Lying-In: A History of Childbirth in America.* New York: Schocken.

Wickler, Wolfgang. 1973. *The Sexual Code: The Social Behavior of Animals and Men*. Garden City, N.Y.: Doubleday, Anchor Books.

Wilkins, M. H. F., A. R. Stokes, and H. R. Wilson. 1953. "Molecular Structure of Deoxypentose Nucleic Acid." *Nature* 171:738–740.

Williams, George C. 1975. *Sex and Evolution*. Princeton, N.J.: Princeton University Press.

Wilson, Edward O. 1975. *Sociobiology: The New Synthesis*. Cambridge: Harvard University Press.

———. 1978. *On Human Nature*. Cambridge: Harvard University Press.

Woolf, Virginia. 1928. *Orlando*. New York: Harcourt Brace and Co., Inc. (Signet Classics Edition, 1960).

———. [1931] 1977. Speech of January 21, 1931, in *The Pargiters: The Novel-Essay Portion of* The Years. Edited by Mitchell A. Leaska. New York: The New York Public Library and Reader Books.

Wright, Barbara. 1979. "Causality in Biological Systems." *Trends in Biochemical Sciences* 4:N110–N111.

Wyman, Jeffries. 1965. "The Binding Potential: A Neglected Linkage Concept." *Journal of Molecular Biology* 11:631–644.

Index

abortion: access to, 165; and fetal rights, 156, 173–174; prenatal testing and, 147, 156, 162–163; problems with selective, 197–198; right to, 162, 195–196, 197; technology of, 144
academia, women in, 35–42
Adam's Rib (Herschberger), 102
adaptive traits, 109–110
adoption, 164–165
AFP, 153
aggressiveness, 117
agriculture, 105
AIDS, 211
algae, sexual behaviors of, 98
alleles, 75–76
alpha-feto-protein (AFP), 153
altruism, 117
amniocentesis, 148, 153, 154, 163
analogous traits, 113
androgens, fetal levels of, 28
anencephaly, prenatal testing for, 153, 194–195
Arendt, Hannah, 191
Arrowsmith (Lewis), 56
Atwood, Margaret, 144

backgrounding, 51
bacteriophage, 84
Baer, Erwin, 185
barbiturates, in childbirth, 150
Beard, Mary Ritter, 35
Beauvoir, Simone de, 119
Bernal, J. D., 60

Binding, Rudolf, 188
biodeterminism, and human nature, 108–109
biology: body build and strength, 120–123; in context, 127–128; and human nature, 107–108; meaning of differences in, 128–129; menstruation, 124–127; molecular, 53; rethinking women's, 119–129; self-serving descriptions of, 119–120; as social construct, 119; and work, 123–124
Birke, Lynda, 114
birth, *see* childbearing
birth-control movement, 143
birth-control pill, 166, 210–211
bisexuality, 135
Blackwell, Antoinette Brown, 93
blood-protection laws, 187–188
body build, 120–122
Bohr, Niels, 116
Boston Women's Healthbook Collective, 33, 68
Bragg, Lawrence, 59, 62, 64–65
brain, fetal hormone levels and, 28
breastfeeding, in !Kung, 126
Britain, eugenics of, 181–182
Brown, Louise, 199, 204

Caesarean section: after in vitro fertilization, 205; right to refuse, 157–159
Cannon, Walter B., 78

Carpenter, Edward, 132
cause and effect, 14, 18
cells, differentiation of, 84
Central Dogma of genetics, 78
"characters," 71, 73, 74–75
Chargaff, Erwin, 59, 61
childbearing: and disabilities, 169–171; and fetal rights, 171–175; and fetal therapy, 175–177; health risks of medicalizing, 148–156; question of choice in, 164–165; as social construct, 161–163; trade-offs of scientific progress for, 165–168
childbed fever, 149–150
Chodorow, Nancy, 133–134
choice: in childbearing, 164–165; power, risks, and, 144–146; prenatal technology and, 169, 191–197; procreative, 141–143
chorionic villus sampling (CVS), 148, 154, 163
chromosomes, 71
Clarke, Edward H., 37–39
clitoris, 131
coeducation, 37–39
colleges, women's, 45, 46
Commoner, Barry, 77
competitiveness, in science, 64
complementary models, of human nature, 116–117
condoms, 166
consent, informed, 152
context stripping, 29–31
contraception, 144, 166
convergent evolution, 113
Crick, Francis, 49, 53–55, 57, 59–65, 78
CVS, 148, 154, 163

Darwin, Charles, 71, 87–97
Davenport, Charles, 183
Dawkins, Richard, 111–112
Delbrück, Max, 49, 53–54
DeLee, Joseph, 150
D.E.S., in pregnancy, 151

Descent of Man, The (Darwin), 94–97
development, sexual, 133–134
developmental genetics, 81
developmental mutants, 81
De Vries, Hugo, 71
dialectical models, of human nature, 114–116
diethyl stilbestrol (D.E.S.), in pregnancy, 151
differentiation, 84
Dinnerstein, Dorothy, 133–134
disabilities: eugenics and, 179–180, 186–187; prenatal technologies and, 164, 169–171, 179–180, 192–195
division of labor, 104–105, 117
DNA: double-helical model of, 48–66; as information, 77–79; and patterns of inheritance, 75–76, 80–81
DNA sequencing, 83–86
domestic science movement, 32
dominance hierarchies, 117
dominant character, 74–75
Donohue, Jerry, 60, 62
Double Helix, The (Watson), 49, 54–55, 58, 61, 64
Down syndrome, 167, 169, 189
drosophila, genetics of, 84–85
Duffey, Eliza B., 39

education: of fact makers, 23–24; women in, 35–42
Edwards, Robert, 199, 204
Ehrhardt, Anke, 136
Eiseley, Loren, 88
Ellis, Havelock, 132
embryo research, 201, 205–208, 210
employment, see work
Engels, Friedrich, 90, 97–98
entropy, 15
environment: genetics and, 114–117; and sex differences, 137–138
episiotomy, 150
equal pay, 25
eugenics: in Britain and United

States, 180–184; genetic engineering and, 81–82; in Germany, 180–181, 184–191; positive and negative, 182–183; prenatal testing and, 191–197; procreative choice and, 143
euthanasia, 188–190
evolution: convergent, 113; Darwin's theory of, 87–92; of man, 102–106; other forms of scientific sexism, 98–102; sexual selection in, 92–98
exchange value, 19–20

fact(s): backgrounding and foregrounding of, 51; in connection of science to nature, 15–16, 50–51; making of, 22–25
"factors," 71–72, 75
fertility drugs, 200, 202
fertilization, 102. *See also* in vitro fertilization
fetal rights, 156–157, 171–175
fetal therapy, 155–157, 175–177
fetoscopy, 148, 155
Firestone, Shulamith, 42
Fischer, Eugen, 185–186
Fisher, R. A., 75
Fletcher, Joseph, 70, 82, 171–172
forceps, 150, 151
foregrounding, 51
Franklin, Rosalind, 49, 55, 57–65
Freire, Paulo, 22, 29
Freud, Sigmund, 133
Fried, Barbara, 136
Frigoletto, Frederic, 176
Frisch, Rose, 125
Frisch, Wilhelm, 187
fruit flies, genetics of, 84–85

Galton, Francis, 70–71, 181
Gamble, Eliza Burt, 87, 93
gathering, 104–105
Geist, Valerius, 99–100
gender, sex vs., 136–137
gene(s), 71–72; alleles of, 75–76;

and cellular differentiation, 84; functioning of, 79–83; and patterns of inheritance, 75–76; and specificity, 72–73, 77
gene mutation, 72, 76, 81
gene therapy, 82, 83, 206–207
genetic engineering, 70, 81–83
genetics, 70–73; developmental, 81; DNA as information, 77–79; and environment, 114–117; functioning of genes, 79–83; Mendel's laws, 73–77; sequencing of human genome, 83–86
genome, sequencing of human, 83–86
genotype, 72
Germany: embryo research in, 210; prenatal testing in, 180–181, 196; racial hygiene in, 181–184
germ-line gene therapy, 207
Glass, Bentley, 171, 189
Goldzieher, Joseph, 210–211
Gombrich, Ernst, 50
Gosling, R. G., 57, 58, 60, 63
Graham, Loren, 10
Guthrie, Woody, 193
Gwaltney, John Langston, 68

Haldane, J. B. S., 72
Handmaid's Tale, The (Atwood), 144
Haraway, Donna, 77–78
Harrison, Michael, 175–176
Heide, Wilma Scott, 40–41
height, 121–122
hemoglobin, patterns of inheritance for, 76
hereditarianism, 70, 73
hereditary disorders: eugenics and, 182–183, 186–187; prenatal testing for, 192–194
Herrnstein, Richard, 114
Herschberger, Ruth, 102
heterosexuals, 132, 134
Higginson, Thomas Wentworth, 39
HIV, 211
Hoche, Alfred, 188

holism, and human nature, 108
Hollingsworth, Leta S., 159–160
home birth, 149
homologous traits, 113
homosexuals, 132–133, 134, 135
Howells, William, 104
Human Genome Initiative, 65,
    83–86, 146
human immunodeficiency virus
    (HIV), 211
human nature: biodeterministic
    models of, 108–109; biological
    models of, 107–108; complemen-
    tary models of, 116–117; dialec-
    tical models of, 114–116; holistic
    models of, 108; interactive models
    of, 113–114; problems with con-
    cept of, 117–118; reductionist
    models of, 108; sociobiological
    models of, 109–113
hunting, 103–105, 106
Huntington's disease, prenatal test-
    ing for, 192–194
Huxley, Julian, 181–182
hybridization, 74–75
hysterectomy, see sterilization

Immigration Restriction Act,
    183–184
incest, 118
induced labor, 150
infertility, 202–203
information, DNA as, 77–79
informed consent, 152
inheritance patterns, 75–76
interactive models, of human na-
    ture, 113–114
intermarriage, 187–188
in vitro fertilization: and embryo re-
    search, 205–208; first successful,
    199; health issues with, 201–202;
    indications for, 203; nature of
    technology for, 144, 200–201;
    prenatal testing and, 167–168; so-
    cial considerations with, 202–
    205; success rate for, 203–204
Irvine, William, 89–90

Jensen, Arthur, 114
Johannsen, W., 71, 72

Kalckar, Hermann, 49
Kamin, Leon J., 114
Keller, Evelyn Fox, 43
Kelsey, Frances E., 151
Kendrew, J. C., 62
Kennedy, Foster, 188–189
Kinsey, Alfred, 134
Klug, Aron, 58
knowledge, 8, 32
Konner, Melvin, 126
!Kung: childbirth in, 161; menstrua-
    tion in, 125–126

labor, 150–151
Lamarck, Jean de, 89
language: evolution of, 106; of sci-
    ence, 12–14
laparoscopy, 200
Laughlin, Harry W., 183, 184, 186
lead exposure, 26, 177–178
Leavitt, Judith Walzer, 149, 166
LeGuin, Ursula, 66
Lenz, Fritz, 185
Lewis, Sinclair, 56
Lewontin, Richard, 114
Linnaeus, Carolus, 88
Lorenz, Konrad, 98
Luria, Salvador E., 53–54, 63
Lyell, Charles, 89

McLaren, Anne, 206
Malthus, Thomas, 73, 91
manual labor, vs. mental labor, 24
marathon runners, 123
marriage, 97–98
Martin, Emily, 68–69
Marx, Karl, 90
maternal serum alpha-feto-protein
    (MSAFP) levels, 153
Mead, Margaret, 117, 137
medical schools, 37–38
Mendel, Agata, 73, 80–81
Mendel, Gregor, 71, 73–77
Mengele, Joseph, 191

menstruation, 124–127; and coeducation, 38–39; and growth pattern, 121–122
mental labor, vs. manual labor, 24
Messer-Davidow, Ellen, 136, 138, 140
midwives, 148–149
Miller, Alice, 139–140
Millikan, Robert, 46
miscegenation, 187–188
molecular biology, 53
Money, John, 136
monogamy, 97–98
Moore, Withers, 39–40
Morgan, Elaine, 94–95
Morgan, Thomas Hunt, 71, 80
Morison, Robert, 10
morphine, in childbirth, 150
MSAFP levels, 153
Muller, H. J., 70, 82
Müller-Hill, Benno, 191
mutants, developmental, 81
mutation, of genes, 72, 76, 81

Nägeli, Karl, 71
nature: interactive, dialectical, and complementary models of, 113–117; laws of, 7–8; and nurture, 114–117; and science, 14–16, 51; woman's, 25–28. See also human nature
Needham, Joseph, 7
negative eugenics, 182–183
neural-tube defects (NTDs), prenatal testing for, 153, 194–195
norms of reaction, 114
Nürnberg laws, 187–188
nursing, in !Kung, 126

Oakley, Kenneth, 103
objectivity, 29–31
occupational hazards, 26, 124
Olby, Robert, 57, 65
Origin of Species, The (Darwin), 91, 93–94

parental investment, 101–102, 110–111

Pauling, Linus, 54, 59, 60, 63
peer review, 23–24
Perry, Ruth, 166
Perutz, Max F., 60, 63
phage, 84
phenotype, 72
Plötz, Alfred, 184
Polanyi, Michael, 65
polygenic traits, 79–80
positive eugenics, 182
pregnancy: fetal rights in, 156–157, 171–175; fetal therapy in, 155–157, 175–177; health risks of medicalizing, 148–156; legal rights of, 156–159; occupational hazards during, 26; of older women, 166–167; trade-offs of scientific progress for, 165–168
prenatal diagnosis: and abortion, 147–148, 162–163; and choice, 169, 191–197; and disabilities, 164, 169–171, 179–180; and eugenics, 191–197; false positives in, 195; health risks of, 153–155; of Huntington's disease, 192–194; and in vitro fertilization, 167–168; methods of, 148, 153–156; and neural-tube defects, 194–195; for older women, 167; and retinitis pigmentosa, 194; right to refuse, 156–157, 173–174; and risk, 144–145, 153–156; and sex selection, 169
procreation, sexuality and, 130–131
procreative choice, 141–143
Proctor, Robert, 186, 189, 190
progestin, in pregnancy, 151
psychoanalytic theory, 139–140
puerperal fever, 149–150

quantitative traits, 79
quantum physics, 116–117

racial hygiene, 184–191
Randall, John, 55, 57, 63
rape, 112
recessive character, 74–75

reductionism, 4–5; and interest in DNA, 52–53, 54, 65; and theories of human nature, 108
Reed, Charles B., 150
reification, 12–14, 51
reproductive fitness, 110
retinitis pigmentosa, prenatal testing for, 194
Rh antigen: genetic basis of, 76–77; prenatal diagnosis of, 147
Rich, Adrienne, 150
Robertson, John, 173–174
Rockefeller, John D., 92
Rocky Mountain bighorn sheep, sexual stereotyping in, 98–100
Rose, Steven, 114
Rossiter, Margaret, 43–44, 46
Rothman, Barbara Katz, 145, 164
Rüdin, Ernst, 186
Russell, Bertrand, 90

Sanger, Margaret, 143
Sayre, Anne, 50, 58
schizophrenia, prenatal testing for, 170–171
Schrödinger, Erwin, 53
science: changing of, 20–21; competitiveness in, 64; language of, 12–14; and nature, 14–16, 51; objectivity of, 9; roles of women in, 31–34; women as objects vs. makers of, 16–19
scientific inquiry, limits of, 10–12
scientific work, as value vs. exchange value of, 19–20
scientists, women as, 43–47
scopolamine, in childbirth, 150
Search, The (Snow), 56–57
selfishness, 112, 117
sex, vs. gender, 136–137
Sex and Evolution (Williams), 100–101
sex differences: in body build and strength, 120–123; construction of, 136–140; environment and, 137–138; inborn, 137; meaning of, 128–129; in work, 123–124

sex roles, 137
sex selection, 169
sexual development, theories of, 133–134
sexual drives, 139
sexuality: childhood, 131; and procreation, 130–131; psychoanalytic view of, 139–140; social construction of, 130–135; toward a nondeterministic model of, 134–135
sexual scripts, 132–133
sexual selection: contemporary views of, 98–102; Darwin's theory of, 92–98
Shaw, Margery, 172–173
Shriner, Thomas, 157–158
sin, sexuality and, 130
Sinsheimer, Robert, 11
Snow, C. P., 56
social class, and work, 27
Social Darwinism, 92
sociobiological models, of human nature, 109–113
somatic gene therapy, 207
specificity, genes and, 72–73, 77
speech, evolution of, 106
Spencer, Herbert, 91, 92
spina bifida, prenatal testing for, 153, 194–195
spinal anesthesia, in childbirth, 151
Sponer, Hertha, 46
sports: and menstruation, 125; and strength, 123
Stadler, L. J., 72
Stent, Gunther, 49
Steptoe, Patrick, 199, 204
sterility, coeducation and, 39
sterilization: and employment, 26; in Germany, 185, 186, 187, 188; laws on, 183; and questions of choice, 165
Stokes, A. R., 58, 60, 63
strength, 122–123
subjectivity, 29–31
survival of the fittest, 91, 101
syphilis experiment, 210

Tanner, Nancy, 104
Terman, Lewis, 182
territoriality, 117
thalidomide, 151
Thomas, Martha Carey, 39
Todd, A. R., 62
tokenism, 45–46
tool making, 102–103, 106
traits: adaptive, 109–110; analogous vs. homologous, 113; DNA and, 78–79; in Mendelian genetics, 71, 73, 74–75; polygenic, 79–80; quantitative, 79
transformationism, 115
twilight sleep, 150–151

ultrasound: and fetal therapy, 175, 176; health risks of, 153–154; and in vitro fertilization, 202; routine use of, 162
unique phenomena, 15
United States, eugenics in, 182–184
use value, 19–20

vagina, 131
Verschuer, Otmar von, 187, 191

Wallace, Alfred Russel, 90, 91
Watson, James, 49, 52–65, 83, 146
Watts, Alan, 50
weight, 122
Weismann, August, 71
Wickler, Wolfgang, 98–99
Wilkins, Maurice, 49, 54–55, 57–58, 60–63, 65
Williams, George, 100–102
Wilson, E. O., 13, 29, 118
Wilson, H. R., 60
women: in academia, 35–42; nature of, 25–28; as objects vs. makers of science, 16–19; roles in science, 31–34; as scientists, 43–47
Woolf, Virginia, 9, 42
work: biology and, 123–124; definition of, 127–128; discrimination in, 25–27, 45–46; equal pay for, 25; hazards of, 26, 124; manual vs. mental labor, 24
Worthman, Carol, 126
Wyman, Jeffries, 12–13

Zihlman, Adrienne, 104